SpringerBriefs in Earth Sciences

More information about this series at http://www.springer.com/series/8897

Eric Bird · Nick Lewis

Beach Renourishment

Springer

Eric Bird
Department of Geography
University of Melbourne
Melbourne, VIC
Australia

Nick Lewis
Royal HaskoningDHV
Sydney, NSW
Australia

ISSN 2191-5369 ISSN 2191-5377 (electronic)
ISBN 978-3-319-09727-5 ISBN 978-3-319-09728-2 (eBook)
DOI 10.1007/978-3-319-09728-2

Library of Congress Control Number: 2014950362

Springer Cham Heidelberg New York Dordrecht London

Printed on acid-free paper

Springer is part of Springer Science+Business Media (www.springer.com)

Preface

During the past century many of the world's beaches have been depleted by erosion. Some beaches, particularly in the United States, Western Europe and Australia, have been renourished by dumping sand or gravel on the shore—the aim being to restore and maintain a beach that will protect the coast from erosion by storm damage and prevent flooding of the hinterland, while providing an improved area for seaside recreation and habitat for wildlife.

Some of the first documented *Beach Renourishment* projects were undertaken in the early 1900s on the east coast of the United States. Several countries have since renourished beaches, particularly during the past few decades. These are reviewed, and experience from various beach renourishment projects from around the world used to discuss principles and practices.

Beach Renourishment will be most effective when those concerned understand how the beach is changing and why. Accordingly, sources of beach sediment and the causes and typical responses to beach erosion are discussed. *Beach Renourishment Principles* dealt with include the need for preliminary investigations, sources of sediment for beach renourishment, methods of beach renourishment, design considerations, techniques for monitoring changes, assessment of performance and modelling and planning considerations.

The text provides researchers, students, engineers, planners and managers with an overview of key guiding principles of beach renourishment. The coverage is necessarily selective and somewhat personal to the authors' respective geographical backgrounds.

A list of references provide a guide to more detailed information, including many technical guides and manuals, along with further details on the specific examples used.

Eric Bird is a Principal Fellow in the Department of Geography, University of Melbourne and the author of several books including *Beach Management* (1996), *Coastal Geomorphology: An Introduction* (1999, 2008), *Submerging Coasts* (1993) and *Encyclopedia of the World's Coastal Landforms* (2010).

Nick Lewis is a Senior Scientist and consultant at Royal HaskoningDHV with a background in environmental management and engineering. With a focus on the practical application of geomorphology, Nick's international experience has been associated with port and harbour development, environmental management and restoration, coastal and flood risk management, offshore resources, transport infrastructure and land development.

Contents

Chapter 1
Introduction

Abstract This chapter provides an introduction to concepts and sets the context for the topic of beach renourishment. Issues with definitions and terminology are discussed and a number of technical manuals and guides introduced.

Beaches can be defined as accumulations of generally loose, unconsolidated sediment on the shore. Some beaches are long and almost straight or gently curved; others are shorter, and include sharply curved 'pocket beaches' in bays or coves between rocky headlands. Beaches fringe about 40 % of the world's coastline, the remainder being partly rocky, partly marshy or muddy, and partly artificial (Bird 2008). Many are exposed to the ocean or stormy seas, but others are sheltered in bays or behind islands or reefs. Beach systems deal with the interactions between beaches and the processes (wind, wave, tides and currents) that work on them. Most beaches consist of sand (sediment particles with grain size between 0.2 and 2.0 mm), but some contain particles with larger diameters, such as granules (2–4 mm), pebbles (4–64 mm) and cobbles (64–256 mm). Beaches consisting entirely of sediment coarser than sand are termed gravel or shingle beaches.

During the past century many of the world's beaches have been depleted by erosion (Bird 1985). Where beaches are eroding, this is likely to continue and even increase due to the predicted global sea level rise and an increase in storm frequency and intensity associated with global climate alterations (Zhang et al. 2004; IPCC 2013). Some beaches, particularly in the United States, Western Europe and Australia, have been restored by dumping sand or gravel on the shore. Sand or gravel, brought from inland, alongshore or offshore sources, has been deposited mechanically or hydraulically to form a beach that is built higher and wider than the depleted beach (US Army Corps of Engineers 1984), the aim being to restore and maintain a beach that will protect the coast from erosion by storm damage (hurricanes, typhoons or tropical cyclones), and also prevent flooding of the hinterland while providing an improved area for seaside recreation and habitat for

E. Bird and N. Lewis, *Beach Renourishment*, SpringerBriefs in Earth Sciences,
DOI 10.1007/978-3-319-09728-2_1

wildlife. Beach renourishment (also termed replenishment, feeding, restoration, recharge, reconstruction, or fill) is an increasingly used method of coastal management, mainly because it preserves the aesthetic and recreational values of protected beaches by replicating the protective characteristics of natural beach and dune systems.

Beach renourishment is artificial in the sense that the sediment has been brought to the shore by engineers. The term beach renourishment is appropriate where an existing beach has been maintained or extended by deposition of suitable sediment. The term artificial beach should be restricted to situations where there was previously no natural beach, as at Praia da Rocha in Portugal (Psuty and Moreira 1990; Psuty et al. 1992). On some coasts artificial beaches have been inserted along the shore in front of sea walls. On some coasts, notably in Singapore and Malaysia, sediment has been deposited to form new land extending out from the natural coastline, and beaches may be formed along the seaward boundary of such reclaimed land.

The use of beach renourishment as a standard method of coastal protection is a fairly recent phenomenon. Some of the first documented projects were undertaken in the early part of the 1900s in the United States, as at San Pedro in southern California in 1919 (Herron 1980) and at Coney Island, New York in 1922. Since then the use of beach renourishment has increased, particularly in the last three decades, and has come into use in many other countries, becoming a globally adopted practice.

The term renourishment is an expression preferred here as an alternative to nourishment, on two counts. Firstly, beach renourishment is often an on-going processes of periodic introductions of sediment to maintain a beach, rather than a one-off construction. For example, on Upham Beach, Florida beach renourishment has taken place to offset continuing erosion in 1975, 1980, 1986, 1991 and 2000 (Elko et al. 2005). Secondly renourishment reflects the fact that the artificial placement of beach material is (re)nourishing a beach previously nourished naturally, prior to erosion.

Several renourishment methods include those that add to the existing beach sediment budget and those that recycle sediment within a beach or coastal system. Renourishment methods can also be categorised according to the location on the beach where renourishment material is placed. These include shoreface renourishment where submerged nearshore bars are created (either to serve as a source of sediment to be transported shoreward, or to dissipate wave energy), profile renourishment (which can include the backshore dune), and beach renourishment where sediment is placed mainly on the subaerial part of the beach. The latter method has a visible effect that is readily perceived by beach users and residents, and is the most widely used method of renourishment in the world (Finkl and Walker 2004).

Beach renourishment has been most common on marine beaches where the aim has been to counter erosion and protect coastal property (Fig. 1.1). Beaches have also been formed on the shores of some estuaries to provide recreational areas, for example on the Elbe Estuary in Hamburg (Fig. 1.2), in Delaware Bay, United States (Jackson et al. 2010) and beside the Thames at Tower Bridge in London.

Fig. 1.1 The ocean-facing beaches of Australia's Gold Coast in SE Queensland, renourished over the last four decades with sediment pumped northward past the mouth of the Tweed River in northern New South Wales, as well as local placement. Beach renourishment helps to protect coastal property from erosion and flooding, as well as providing valuable recreation space. © Nick Lewis

The concept of beach renourishment expects that the placed sediment, along with the existing beach material, will adapt its shape to the changing wave and tidal conditions and dissipating wave energy. The topic of beach renourishment has a substantial and rapidly growing geomorphological, engineering and environmental literature. Major works on the technical aspects of beach renourishment include the Shore Protection Manual produced by the US Army Corps of Engineers (1984), the Delft Hydraulics Laboratory (1987) manual on beach renourishment, the German Empfehlungen für Küstenschutzwerke (EAK 1993), the American book on Beach Nourishment and Protection (NRC 1995), the Coastal Engineering Manual produced by the United States Army Corps of Engineers (USACE 2002), the textbook Beach Nourishment: Theory and Practice (Dean 2002) and the British Construction Industry Research and Information Association (CIRIA) Beach Management Manual (CIRIA 1996, 2010).

Most renourishment projects have taken place on coasts where the natural beach has been depleted by erosion. Beach renourishment has been used at seaside resorts where erosion had become a problem, in order to restore the beach for recreational use. However, some beaches have been inserted primarily as a means of protecting the coastline by absorbing wave energy and so preventing further cliff

Fig. 1.2 The artificial beach emplaced beside the River Elbe in Hamburg. © John Marquet

erosion or damage to coastal property. A renourished beach may be used as well as, or instead of, hard protective structures such as sea walls or boulder ramparts.

The principles, practices and problems of beach renourishment, based on a review of projects that have been documented in published or readily available literature from various parts of the world will now be introduced and discussed. It is first sensible to consider the causes of beach erosion (the need for renourishment) and responses to beach erosion.

There have been failures, and local conditions must always be taken into account when recommending, planning and carrying out a beach renourishment project. This account is intended to provide background for those concerned with coastal planning and management.

References

Bird ECF (1985) Coastline changes: a global review. Wiley, Chichester

Bird ECF (2008) Coastal geomorphology: an introduction, 2nd edn. Wiley, Chichester

CIRIA (1996) Beach recharge materials—demands and resources. In: Humphreys B, Coates T, Watkiss M, Harrison D (eds) Report. Construction Industry Research and Information Association, London, p 154

CIRIA (2010) Beach management manual (report no. C685). CIRIA, London

Dean RG (2002) Beach nourishment: theory and practice. Advanced series on ocean engineering, vol 18. World Scientific Publishing, Singapore, p 399
Delft Hydraulics Laboratory (1987) Manual on artificial beach nourishment. Centre for Civil Engineering Research, Codes and Specifications, Gouda
Elko NA, Holman RA, Gelfenbaum G (2005) Quantifying the rapid evolution of a nourishment project with video imagery. J Coast Res 21:633–645
Empfehlungen fur die Ausfuhrung von Küstenschutzwerke—Recommendations for the design of coastal structures (EAK) (1993) Die Küste. Heft 55, Heide, Germany (in German)
Finkl CW, Walker HJ (2004) Beach nourishment. In: Schwartz M (ed) The encyclopedia of coastal science. Kluwer, Dordrecht, pp 147–161
Herron WJ (1980) Artificial beaches in Southern California. Shore Beach 48:3–12
Intergovernmental Panel on Climate Change (IPCC) (2013) Climate change 2013: the physical science basis. Contribution of working group I to the fifth assessment report of the intergovernmental panel on climate change, Geneva
Jackson NL, Nordstrom KF, Saini S, Smith DR (2010) Effects of nourishment on the form and function of an estuarine beach. Ecol Eng 36:1709–1718
National Research Council (NRC) (1995) Beach nourishment and protection. National Academy Press, Washington, D.C
Psuty NP, Moreira ME (1990) Nourishment of a cliffed coastline, Praia da Rocha, the Algarve, Portugal. J Coast Res 6:21–32
Psuty NP, Eugine M, Moreira SA (1992) Characteristics and longevity of beach nourishment at Praia da Rocha, Portugal. J Coast Res 8:660–676
US Army Corps of Engineers (1984) Shore protection manual (2 volumes), 4th edn
US Army Corps of Engineers (USACE) (2002) Coastal engineering manual. Washington, DC
Zhang K, Douglas BC, Leatherman SP (2004) Global warming and coastal erosion. Clim Change 64:41–58

Chapter 2
Causes of Beach Erosion

Abstract Before renourishing an eroded beach it is necessary to know why it has been eroded and where the sediment has gone: landward, seaward or alongshore. This chapter deals with the causes of beach erosion, including alterations in processes and sediment supply, along with anthropogenic influences.

Before renourishing an eroded beach it is necessary to know why it has been eroded and where the sediment has gone: landward, seaward or alongshore (Fig. 2.1).

Beach erosion is usually marked by the evolution of a concave-upward shore profile, whereas accreting (prograding) beaches typically have convex-upward shore profiles. There is sometimes a receding microcliff (an erosional scarp), a metre or more in height, where an upper convex beach is being undercut as a lower concave profile becomes established (Fig. 2.2). Backshore dunes are often cliffed behind beaches that have been lowered and cut back by erosion, as are backshore terraces, and in both cases vegetated land surfaces are truncated as erosion proceeds.

Erosion can be temporary, reversed by a following period of accretion, or long-term, resulting in a net retreat of the coastline (recession). The retreat of the coastline can be traced by comparing dated sequences of maps and charts, or air and ground photographs. Average annual rates of coastline recession are usually small (a few centimetres each year), but there have been instances of recession rates of more than 40 m per year, as on beaches fringing rapidly eroding delta shores, or on beaches fronting relatively less resistant cliffs, such as those comprising glacial deposits for example.

Beach erosion can be caused by natural or anthropogenic alterations to the sediment budget (including both the sources and sinks of beach sediment) or the processes that work on them. The main causes of beach erosion are as follows:

1. Reduction in sediment supply from eroding cliffs
2. Reduction of fluvial sediment supply to the coast
3. Reduction of sediment supply from the sea floor

E. Bird and N. Lewis, *Beach Renourishment*, SpringerBriefs in Earth Sciences,
DOI 10.1007/978-3-319-09728-2_2

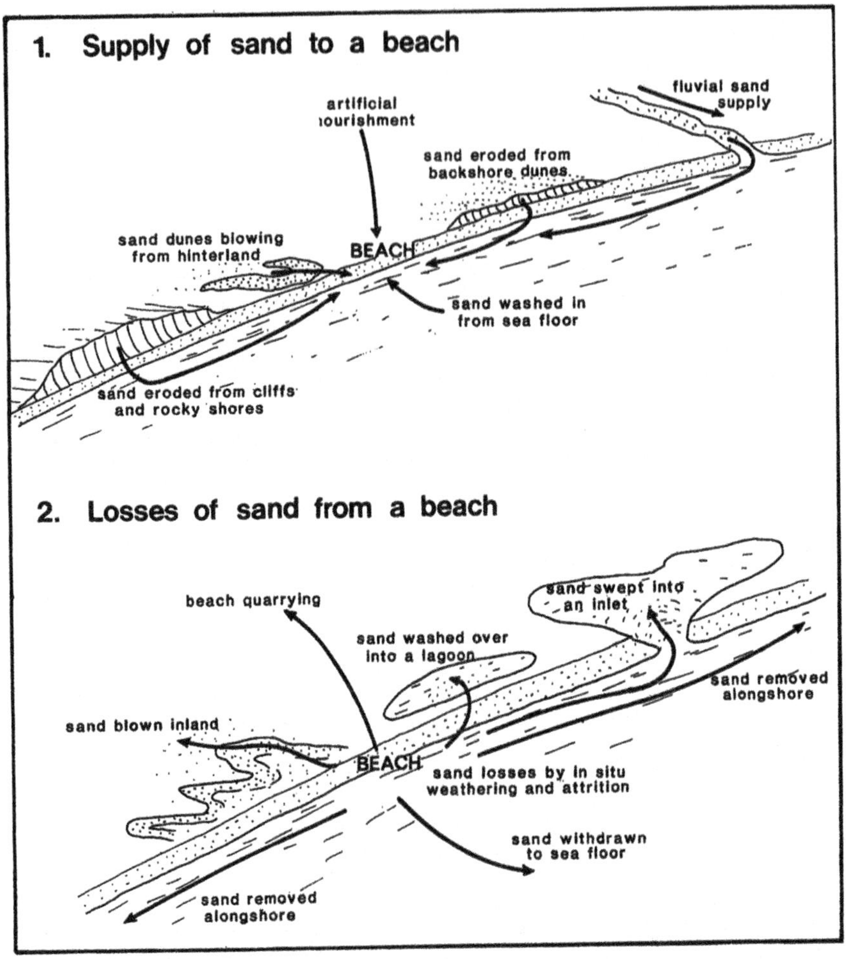

Fig. 2.1 The various ways in which sand can be supplied by and removed from a beach. © Geostudies

4. Reduction of sand supply from inland dunes
5. Submergence and increased wave attack
6. Increased wave energy because of increased storminess
7. Losses of beach sediment alongshore
8. A change in the angle of incidence of waves
9. Interception of longshore drift by breakwaters
10. Increased losses of beach sediment to the backshore
11. Beach weathering, including attrition of beach sediment
12. A rise in the beach water table
13. Removal of beach sediment by runoff

Fig. 2.2 Microcliff (erosional scarp) on Collaroy-Narrabeen Beach, NSW, Australia, cut by a number of late summer storms in 2013. © Nick Lewis

14. Increased scour by wave reflection from an artificial structure
15. Extraction of sand and shingle from the beach

Erosion on a particular beach is generally due to more than one of these causes, although one cause is often dominant.

2.1 Reduction in Sediment Supply from Eroding Cliffs

A common cause of beach erosion is the reduction of the supply of sand or gravel from erosion of nearby cliffs. Stabilisation of a cliff to halt erosion usually takes the form of building a solid wall or boulder rampart along the base of a cliff to prevent wave attack. Beach erosion also occurs when the sediment supply from an eroding cliff by runoff, seepage and slumping is reduced by inserting drains or introducing vegetation or a geotextile carpet.

As cliffs are stabilised their sediment yield to the shore diminishes, and may cease altogether. Beach erosion ensues as sediment lost offshore (mainly during storms) or alongshore (when waves arriving at an angle to the shore generate longshore drift) is no longer replenished from an eroding cliff. This happened on the

Fig. 2.3 Canford Cliff, Bournemouth, Dorset cut in Tertiary sandstones capped by Pleistocene gravel, as it was in 1950. Sand and gravel eroded from the cliff by runoff and marine erosion was then being delivered to adjacent beaches. See Fig. 2.4. © Geostudies

Fig. 2.4 Canford Cliff after the construction of a concrete esplanade along the base of the cliff. This halted marine erosion, but also reduced the supply of sand and gravel to adjacent beaches, which became depleted by marine erosion. © Geostudies

coast of Bournemouth in southern England after a concrete promenade built along the base of eroding cliffs cut in soft sandstone and gravel (Figs. 2.3, 2.4) cut off the supply of sand and gravel to the beach, which then gradually diminished. The promenade was built partly for the benefit of seaside holidaymakers, but it was also intended to halt cliff recession and preserve coastal properties. The coastal slope was artificially landscaped and planted with vegetation, but by the 1970s Bournemouth beach was severely depleted, and it was decided that it should be renourished (Sect. 4.2.7, p. 49).

2.2 Reduction of Fluvial Sediment Supply to the Coast

Beach erosion occurs where beaches that have been supplied with sediment carried down to the coast by rivers are depleted following a reduction in sediment yield to river mouths as a result of reduced runoff. In Southern California diminished river flow during droughts resulted in beach erosion, but the beaches were restored during intervening wet years when the fluvial sediment supply revived (Orme 1985). Reduction of fluvial sediment supply commonly results from the construction of dams to impound water upstream. These intercept fluvial sediment discharge and so cut off the supply of sand and gravel to beaches at and near the river mouth. This leads to the onset of erosion on beaches that were formerly maintained or prograded by the arrival of this fluvial sediment. Erosion develops more quickly, and becomes more severe, where there is strong longshore drift of sediment away from the river mouth.

The best known example of such erosion is on the shores of the Nile delta, where sandy beaches that had been prograding for many centuries as the result of the delivery of sediment to the mouths of Nile distributaries and its distribution by longshore drift became depleted after the construction of dams upstream. Erosion of beaches near the mouths of the Rosetta and Damietta distributaries started soon after barrage construction began in 1902, and became much more rapid and extensive after the completion of the Aswan High Dam in 1964, which impounded Lake Nasser and resulted in large-scale sediment entrapment. During the next few years beach erosion on parts of the deltaic coastline attained annual rates of up to 120 m (Sestini 1992). Some of the sediment removed from these beaches was carried away eastward by longshore drift along the coast towards Port Said, but much has been lost offshore (Lotfy and Frihy 1993).

Similar beach erosion has occurred on the shores of other deltas following dam construction upstream: for example on the Rhône delta in France, the Dnieper and Dniester deltas in the Ukraine, the Citarum delta in Indonesia and the Barron delta in Australia.

Diversion of a river mouth, either naturally or artificially, halts the fluvial sediment supply to the coast and leads to erosion of adjacent beaches, as on the shores of the Cimanuk delta in Java after a distributary changed its course during a flood in 1947, and on the Hwang Ho delta in China after the river mouth was diverted to another part of the coast during a flood in 1852.

Beach erosion near river mouths can follow the dredging of sand from river channels, which occurred on the River Rhine during the Second World War, or the reduction of fluvial sediment supply by soil conservation works in the hinterland, as exemplified by the rivers draining to the Gulf of Taranto in southern Italy. The same effect has been produced where long-continued soil erosion in river catchments has removed unconsolidated surface sediment, exposing extensive areas of bare rock, as in Turkey and Greece, where consequent reductions in fluvial sediment yield have resulted in beach erosion.

2.3 Reduction of Sediment Supply from the Sea Floor

On many coasts beaches were deposited and prograded when sand was swept in from the sea floor by wave action during and since the Late Quaternary (Pleistocene and Holocene) marine transgression. As sea level rose across the continental shelf waves collected sediment that had previously been deposited by rivers or wind action and sediment from weathered rock outcrops and carried it shoreward, and as the Late Quaternary marine transgression came to an end continuing shoreward drift from shoals prograded these beaches, often forming successive backing beach ridges and parallel dunes (Fig. 2.5a–d).

On many of these coasts shoreward drift has come to an end, with the reshaping of the sea floor profile to a transverse concave profile across which wave action no longer moves sediment on to the shore. If there is no compensating input of sediment from other sources (such as cliff erosion or supply from rivers) beach progradation stops, and with continued input of wave energy the transverse nearshore profile migrates landward, so that the beaches are eroded. This explains why many beaches that prograded earlier in Holocene times are now being cut back by erosion, continuing wave action driving the transverse concave landward (Fig. 2.5d). The onset of erosion comes at different times in different places because the development of the concave profile and the cessation of shoreward drift has occurred at various times on various beaches, and has not yet been attained on coasts where there is still shoreward drift from nearshore shoals.

The sequence portrayed in Fig. 2.6 is illustrated on the Ninety Mile Beach in south-eastern Australia, which borders a sandy coast that formerly prograded by accretion of sand supplied from the adjacent floor of Bass Strait, and is now being cut back by marine erosion, except in a sector of continuing accretion alongside breakwaters at Lakes Entrance (Fig. 2.6). There is still plenty of sand in the nearshore area, but the transverse profile has become smooth and concave, and there is no longer shoreward drift of sand to this beach, except at the SW end, sheltered by the granitic upland of Wilsons Promontory, where there are still nearshore sand shoals.

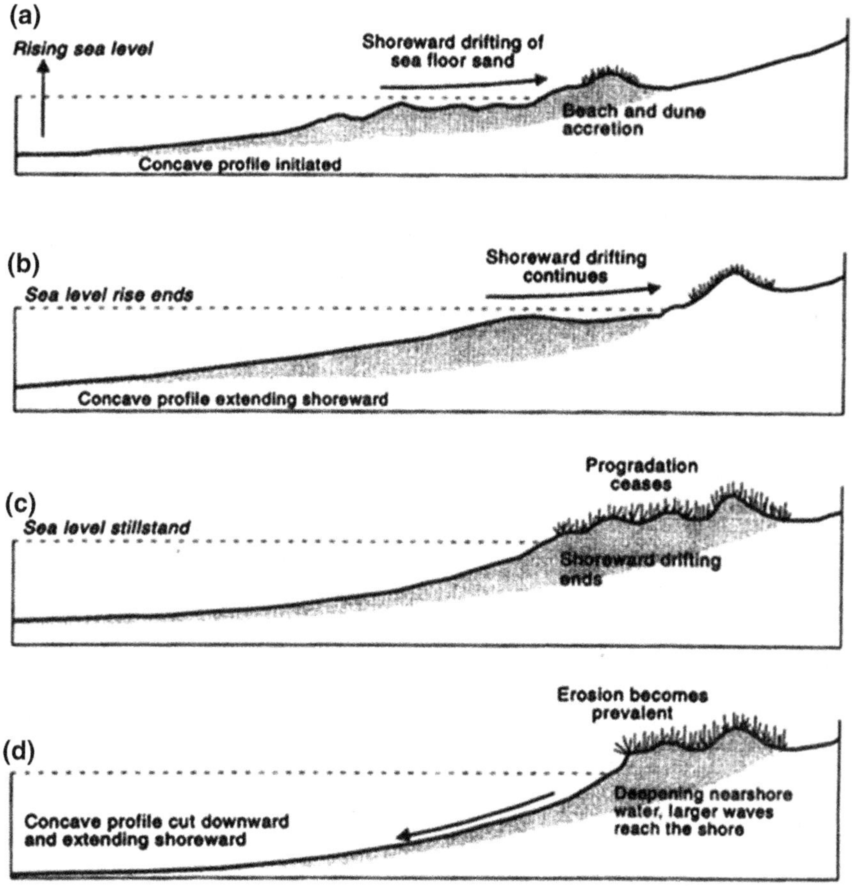

Fig. 2.5 On many coasts sand drifted shoreward from sea floor shoals during the Late Quaternary marine transgression (**a**), and for a time after this transgression ended (**b**), so that beaches prograded. With the attainment of a smooth concave sea floor profile progradation ceased (**c**), and the landward migration of this profile resulted in beach erosion and coastline retreat (**d**). © Geostudies

Sediment supply from the sea floor is also seen where shells, or other biogenic particles derived from sea floor organisms, become calcareous sand and gravel that drifts shoreward to be added to beaches. Several beaches on the west coast of Western Australia are still maintained in this way, supplied with calcareous sand and gravel from disintegrating nearshore reefs of dune calcarenite (which consists of sand, usually mainly calcareous, cemented by precipitated carbonates to form a coherent sandstone).

Such beaches prograde, or are maintained, as long as there is a supply of sea floor sediment, but beach erosion develops if the sediment supply is reduced

Fig. 2.6 Erosion (*arrowed*) of the Ninety Mile Beach, on the outer barrier of the Gippsland Lakes in SE Australia, with accretion (+) on either side of the Lakes Entrance breakwaters. © Geostudies

because of ecological changes such as the destruction of shell fauna by pollution, or because increased growth of seagrasses or other marine vegetation has impeded shoreward drift and trapped sediment offshore.

2.4 Reduction of Sand Supply from Inland Dunes

Some beaches have been supplied with sand blown from dunes spilling from the land on to the shore. If the sand supply runs out, or the backshore dunes become stabilised, either by the natural spread of vegetation, or from the planting of grasses or shrubs, the spraying of bitumen or rubber compounds, or sealing of the dune surface by built structures, these beaches may start to erode. On the south-facing Cape Coast of South Africa, where the prevailing westerly winds are driving dunes over headlands to supply beaches on the lee shore, dune stabilisation has resulted in beach erosion, as at Port Elizabeth.

Where beach erosion is due to the reduction of sand supply from inland dunes this could be the result of successful conservation measures, such as the

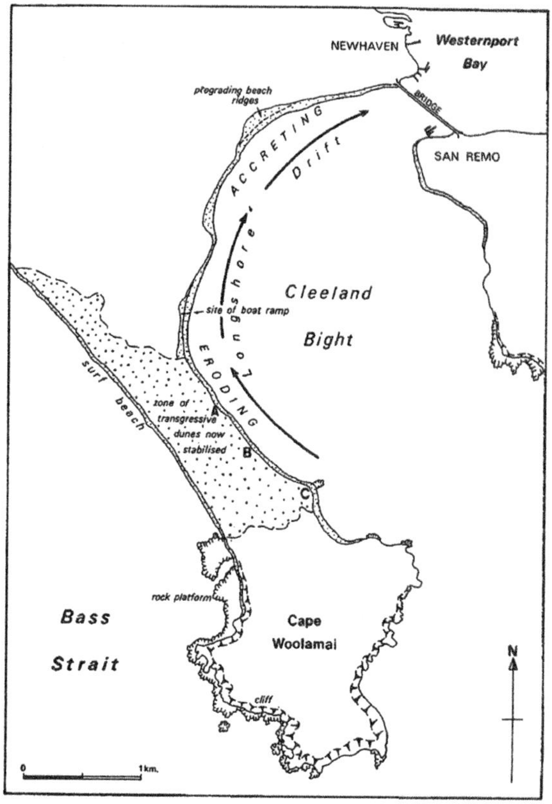

Fig. 2.7 Cleeland Bight on Phillip Island, Australia, showing the northward drift of sand towards Newhaven. When dunes were spilling on to the shore the beach was built upward and outward, but after the dunes were stabilised in the 1980s by planting marram grass this supply was reduced, and beach erosion ensued. © Geostudies

establishment of a vegetation cover on formerly drifting dunes, or the extinction of the dunes that had been drifting to the coast. On Phillip Island, Australia, partial stabilisation of dunes that had been spilling eastward across the Woolamai isthmus to renourish the beach in Cleeland Bight (Fig. 2.7) was followed by beach erosion (Fig. 2.8).

On Balneario Camboriu Beach in southern Brazil intensive urbanization of the coastal zone since the 1960s has resulted in the building of a sea wall, a road and several tall buildings. This process has reduced the amount of sediment exchange between the beach and dune and resulted in erosion during storms and a reduction in beach width (Temme et al. 1997). To minimise the loss of beach area 50,000 m^3 of sand was dredged from the sea floor and placed on an 800 m stretch of the beach in 2002 (Pezzuto et al. 2006). Periodic beach replenishment is required to widen the dry-beach and add height to the berm (Finkl and Walker 2004).

Fig. 2.8 Depleted beach and backshore erosion on the shore of Cleeland Bight, Phillip Island, following the stabilisation of dunes that previously supplied sand to this beach. © Geostudies

2.5 Submergence and Increased Wave Attack

Deepening of nearshore waters allows larger waves to reach the shore and erode the beach, withdrawing sand or gravel to the sea floor. Such deepening occurs briefly during storms, when strong onshore winds raise sea level along the coast and larger-than-usual waves break on the shore, eroding beaches. Longer-term deepening occurs as the result of coastal submergence, produced either by land subsidence, an actual rise of sea level, or some combination of land and sea movement that results in the sea standing higher relative to the land. Larger waves then reach the shore, causing erosion and the re-shaping of the nearshore profile: erosion of the upper beach and transference of sand or gravel from the beach to the adjacent sea floor causes the transverse shore profile to migrate upward and landward.

Beach erosion has become widespread on coasts where the sea has been rising because land subsidence is in progress, as on the Gulf and Atlantic coasts of the United States. Coastal land subsidence resulting from extraction of groundwater has resulted in beach erosion on the northern Adriatic coast of Italy, as at Ravenna, and beaches were cut back suddenly on sectors of the Alaskan coastline that subsided during the 1964 earthquake.

There is evidence from tide gauge records of a sea level rise of 1–2 mm/year during the past few decades, offset on some coasts by equal or greater land uplift, and varying also in relation to the geophysical factors that complicate the surface topography of the oceans. Coastal submergence has been widespread, and may provide at least a partial explanation for the modern prevalence of beach erosion (Bird 1996), although it is often cited as the primary cause of beach erosion (Douglas et al. 2000).

2.6 Increased Wave Energy

As has been noted, beach erosion occurs where wave energy increases (i.e. larger and higher waves approach the shore) because nearshore water is deepened by coastal submergence, due to a rise in sea level or subsidence of coastal land. Nearshore water can also be deepened where a shoal or reef is removed, either by natural erosion or by dredging, or where nearshore seagrass meadows disappear, so that sediment that had been retained is dispersed. On the coast at Benacre Ness, Suffolk, United Kingdom, a sector of beach that had been protected from wave attack by a nearshore shoal began to erode as the shoal moved away alongshore, while in Botany Bay, Australia, beach erosion accelerated at Brighton-le-Sands after the bay floor was dredged to provide sediment for the extension of a runway at Sydney International Airport. In the 1930s deepening of nearshore water following the disappearance of seagrasses (which had retained sediment in the sea floor shoals) led to beach erosion on the shores of Danish islands such as Kyholm (Christiansen et al. 1981). On the Arctic coast of Russia increased beach erosion has been attributed to larger waves arriving as the result of nearshore deepening due to downwarping of the adjacent sea floor.

Some beaches that had been stable or prograding began to erode as the result of an increase in the frequency and severity of storms in coastal waters. A series of storms in quick succession is particularly destructive because the second and subsequent events occur on beaches already reduced to a concave eroded profile. An example of this has been documented from Estonia, where the climate has become stormier during the past few decades, with sea level more frequently raised by storm surges, so that coastline erosion became more rapid and more extensive, notably on the west coast of Saaremaa Island (Orviku et al. 2003).

2.7 Losses of Beach Sediment Alongshore

Beaches are depleted when sand or gravel are carried away by longshore drift (due to the arrival of waves at an angle to the shore), unless these losses are compensated by the arrival of more sediment from updrift. Beach erosion will occur if the losses downdrift exceed the supply from updrift, for example where the source of sediment updrift is a cliff that has been stabilised, or a river where fluvial sediment yield has diminished. Some beaches develop lobes of sand or gravel that migrate along the coast in the predominant direction of longshore drift, and at any particular point there is accretion as each lobe arrives, and erosion as it moves on. At Somers, on the coast of Western Port Bay, Australia, a yacht clubhouse was built on one such lobe in the 1970s, and was threatened by beach erosion that developed as that lobe moved on (Fig. 2.9).

Portsea, on the southern shore of Port Phillip Bay, Australia, has had recurrent beach erosion because of eastward longshore drift (Fig. 2.10). Sand moving along the coast accumulates alongside headlands, and then is swept at intervals round them as a series of migrating lobes. As each lobe arrives at Portsea the beach widens, but as it moves on this beach diminishes. Segments of sea wall have been constructed successively on sectors where beach erosion was severe.

Attempts to blame the erosion of Portsea Beach in 2010 on the dredging of a deeper and straighter entrance to Port Phillip Bay in 2008 failed because of evidence that episodes of beach erosion occurred at several times prior to entrance deepening, and because wave refraction diagrams drawn for configurations before and after dredging showed no change in the pattern of waves or wave energy at Portsea (Cardno 2011). The beach was renourished, but the deposited sand quickly drifted away to the east, and the coast at Portsea was then armoured with sandbags to halt recession (Fig. 2.11).

Fig. 2.9 The yacht club at Somers, Western Port Bay, Australia, was built on a sand lobe (outline: *dotted line*) that had migrated to this position, and then moved on along the coast. A boulder rampart has been inserted to halt erosion here. © Geostudies

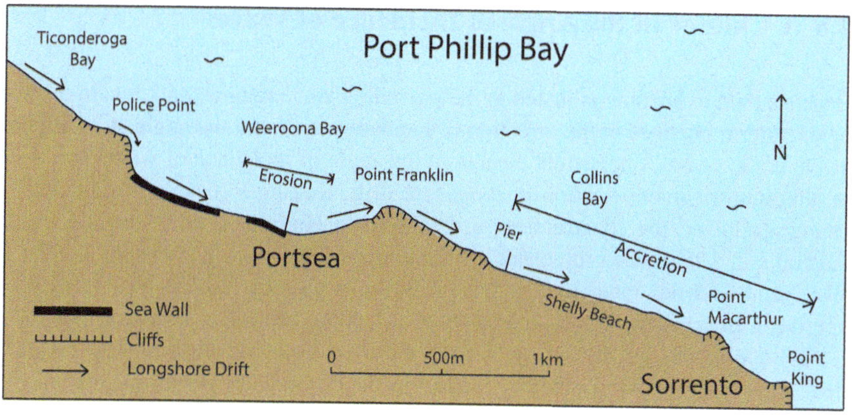

Fig. 2.10 At Portsea, on the north-facing coast of Port Phillip Bay, the predominant direction of longshore drift is from west to east. Lobes of sand form on the shore of Triconderoga Bay, to the w, and drift intermittently round police point into Weeroona Bay. As each sand lobe arrives, the beach widens in Weeroona Bay, but when it moves on the beach is depleated. Successive episodes of beach erosion have resulted in the building of sea walls along the shore. In 2010 there was severe beach erosion at Portsea Pier, with sand drifting away round point Franklin and along the coast to Shelly Beach and Point Macarthur, where there has been beach accretion. A sandbag rampart has been built on the eroded sector (Fig. 2.11), but the beach has not yet revived. © Geostudies

Fig. 2.11 A sandbag rampart marks the site of beach erosion at Portsea, Port Phillip Bay, Australia, in 2013. © Geostudies

2.8 A Change in the Angle of Incidence of Waves

Beaches tend to become adjusted to the prevailing wave regime, and then to respond to short-term changes in the direction of incident waves and the angle at which they arrive at the coast. A persistent change in the angle of incidence of waves can result in alteration or intensification of longshore drift, leading to beach erosion. Such a change followed the construction of the Portland Harbour breakwater in Victoria, Australia in 1957, when the onset of beach erosion on the adjacent coast at Dutton Way resulted from intensified longshore drift in response to the change in wave approach. Beach erosion then spread eastward along the coast of Portland Bay.

Oblique waves generated by passing ships (boat swash) can modify the incident wave regime. This has led to erosion on Point King Beach at Sorrento on Port Phillip Bay, Australia, where the beach was re-shaped as intensified longshore drift depleted the eastern end and led to accretion at the western end (Bird 2011).

Refraction of waves over the ebb shoal at the entrance to John's Pass, West coast of Florida causes a local reversal in longshore sediment transport contributing to severe erosion on Sunshine Beach (Wang et al. 2011).

Wave attack on a beach sector may intensify as a result of the lowering of the beach profile on the neighbouring sector, allowing stronger waves to arrive obliquely, and thus accelerate longshore drift. A beach profile may be lowered as the result of sea wall construction and scour by reflected storm waves on one sector, so that larger oblique waves can then move through the deepened water to

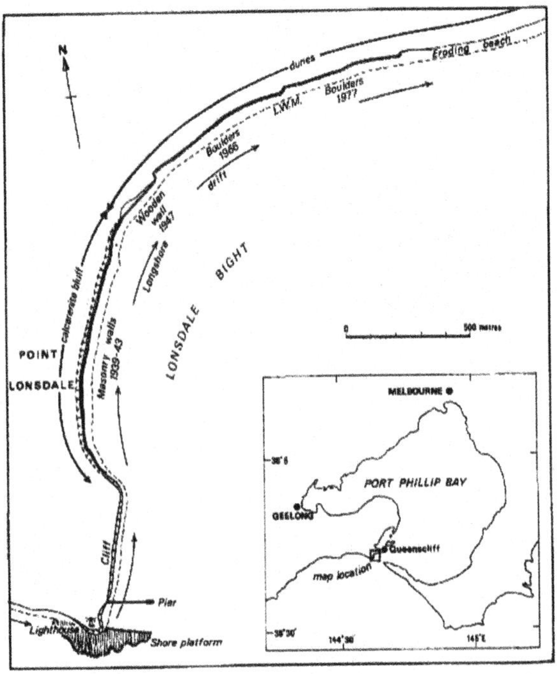

Fig. 2.12 In response to coastal erosion a sea wall was built at Point Lonsdale in 1900. This resulted in beach depletion by reflected waves (cf Fig. 2.15), and also the lowering of the beach to the north, followed by accelerated erosion there. The sea wall was extended northward in the 1930s, with a similar result, and further set-back extensions were made in 1947, 1966 and 1977. © Geostudies

attack the neighbouring sector, causing beach erosion there. If the sea wall is then extended along the coast to counter this beach erosion, a 'domino sequence' may ensue, with beach erosion beyond each limit of the extended sea wall. This happened at Point Lonsdale in Victoria, Australia, where each new sector of sea wall has been built on a set-back alignment on the eroded shore (Fig. 2.12).

2.9 Interception of Longshore Drift by Breakwaters

Breakwaters have been built to stabilise river mouths or lagoon entrances in order to improve their navigation, or create boat harbours. Where beach sediment is drifting alongshore there is interception on the updrift side of the breakwaters, and beach erosion on the downdrift side as the sediment supply is cut off. The downdrift erosion caused by breakwaters may spread for several tens of kilometres (Bruun 1995) and can prove irreversible (El-Asmar and White 2002). On the east coast of Florida breakwaters have been built to stabilise several tidal entrances through sand barriers, and in each case southward longshore drift has been intercepted to prograde the beach on the northern side, and beach erosion has ensued on the southern side, deprived of a longshore sand supply. The erosion at Upham on the west coast of Florida is caused by a significant deficit in the southward longshore sediment transport, due to the structures at the Blind Pass inlet (Elko and Davis 2006, Elko et al. 2005). Practically, no sand bypasses Blind Pass to reach Upham Beach (Roberts and Wang 2012). Finkl and Esteves (1998) estimated that structures blocking littoral drift accounted for 72 % of Florida's beach erosion.

The seaport at Chennai (formerly Madras) on the south east coast of India underwent massive expansion programmes in the last two decades, which resulted in substantial changes in the geomorphology of the down-drift side of the port (Ramana Murthy et al. 2008). Nearly 400 ha of beach was lost as a result of erosion, which resulted in the construction of a seawall.

Until about a century ago there was a shingle beach, maintained by eastward drift, beneath the Chalk cliffs between Dover and Deal in the United Kingdom but this has almost disappeared as the result of the interception of shingle drifting alongshore by the breakwaters at Dover Harbour, west of which the beach has prograded. In Lyme Bay on the south coast of England the dominant eastward longshore drift of shingle has been intercepted by landslide lobes and rock falls, each of which act as breakwaters, causing beach erosion downdrift.

2.10 Increased Losses of Beach Sediment to the Backshore

Sand is swept from the beach to the backshore by strong onshore winds (Fig. 2.13), or when storms wash beach sediment beyond the back of the beach, or over into lagoons, swales or swamps, or the mouths of rivers. If losses from the

Fig. 2.13 Sand is moving inland from a beach as drifting (transgressive) dunes on the shore of Encounter Bay, South Australia. The lowered backshore is then cut back quickly because of the diminished volume of sand to be removed by wave attack. © Geostudies

beach are not compensated by the arrival of fresh supplies of beach sediment from offshore, alongshore or hinterland sources the beach profile is lowered and the coastline recedes. Overwash during successive storm surges has eroded beaches on sandy barrier islands on the Atlantic coast of the United States. At Rockaway and Long Beach on the New York coast overwash during Hurricane Sandy in 2012 swept large amounts of beach sediment shoreward into residential areas.

2.11 Beach Weathering, Including Attrition of Beach Sediment

Beaches no longer receiving a sediment supply lose volume as the result of weathering, which reduces the size of beach particles and hence the volume of the beach, beach profile and so allowing larger waves to attack the shore and further erode the beach. Chemical weathering includes the decay and removal of ferromagnesian minerals from sediments of volcanic origin and the dissolving of carbonate beach sand grains or limestone gravels in rainwater, stream seepage or sea spray. Physical weathering occurs as the result of agitation of the beach by wave action and consequent gradual attrition of sediment particles.

Four Mile Beach, in North Queensland, Australia, has been eroded because its fluvial sand supply from the adjacent Mowbray River was cut off by coral reef growth. Without natural sand replenishment the beach has been reduced to very fine sand by attrition, and has become low and flat: it is now firm enough to land an aircraft, drive a bus or car, or ride a bicycle. As sediment calibre is reduced the lowered beach has been further eroded as the increasingly fine sediment is removed by winnowing, either landward into backshore dunes or seaward to bars and sea floor deposits.

2.12 A Rise in the Beach Water Table

A wet sandy beach is eroded more rapidly by wave action than a dry one because wet sand is more coherent, and erodes like a soft sandstone, whereas dry sand is disturbed but not removed by wave swash. Field studies conducted by Grant (1984) demonstrated the considerable impact of beach groundwater level on swash sediment transport. Seawater infiltration under a low water table was found to enhance onshore sediment transport, whereas groundwater exfiltration under high water table promoted offshore sediment transport. On Stanwell Park Beach, near Sydney, Australia, beach erosion increased with rises in the level of the beach water table during wet weather, due to the ponding or diversion of river or lagoon outlets, or to increased river or groundwater discharge following land use changes in the hinterland (Bryant 1985). This process has led to the practice of beach dewatering (artificially lowering the beach water table) for combating beach erosion, typically utilising drainage systems (Turner and Leatherman 1997, Loannidis and Karambas 2007).

2.13 Removal of Beach Sediment by Runoff

During periods of heavy rainfall beach erosion can result from runoff, particularly where water flows down a backing cliff or steep slope and beach sediment is swept into the sea. Examples of this are seen during the wet summer season in NW Australia, notably near Cape Leveque, where runoff cuts gullies across the beach and builds fans of sediment into the sea. Beach sediment removal by runoff after rainfall occurs at the mouth of a storm water outfall sited within a beach. On Collaroy-Narrabeen Beach, north of Sydney, Australia (Fig. 2.14) beach sediment has been washed away at a number of storm water outfalls. At Mission Bay and Kohimarama on the east coast of Auckland, New Zealand, sand was sluiced from beaches by storm water from outlets until these were extended seaward, beyond the foot of the beach (Papps and Priestley 2005).

The effects of runoff are stronger on sandy beaches, especially if they are already wet, than on gravel where runoff disappears more quickly by percolation. Increased runoff is often due to urbanisation and the construction of roads and other sealed surfaces from which water runs off quickly, instead of percolating into the subsoil, as it did before these structures were built.

Fig. 2.14 A channel cut through the beach and dune on Collaroy-Narrabeen beach, Sydney, Australia caused by storm water discharging from a drainage outfall. © Nick Lewis

2.14 Increased Scour by Wave Reflection from an Artificial Structure

Waves breaking against a solid shore structure, such as a sea wall built of concrete, stone blocks, boulder ramparts, steel sheeting or timber, are reflected, and generate seaward currents that carry sediment away from the foot of the wall (Fig. 2.15). This reflection scour is prevented as long as a beach is high and wide enough to stop waves reaching and reflecting from the solid structure, but beach erosion occurs rapidly once waves reach the backing wall (Fig. 2.16).

2.15 Extraction of Sand and Shingle from the Beach

Sand or gravel has been quarried from many beaches for use in road and building construction (Fig. 2.17), and the result of lowering the shore profile is to allow larger waves to attack the beach more strongly during storms. Beaches in Jersey and Guernsey were much reduced by the extraction of sand from beaches by the German occupying forces during the Second World War for use to build bunkers

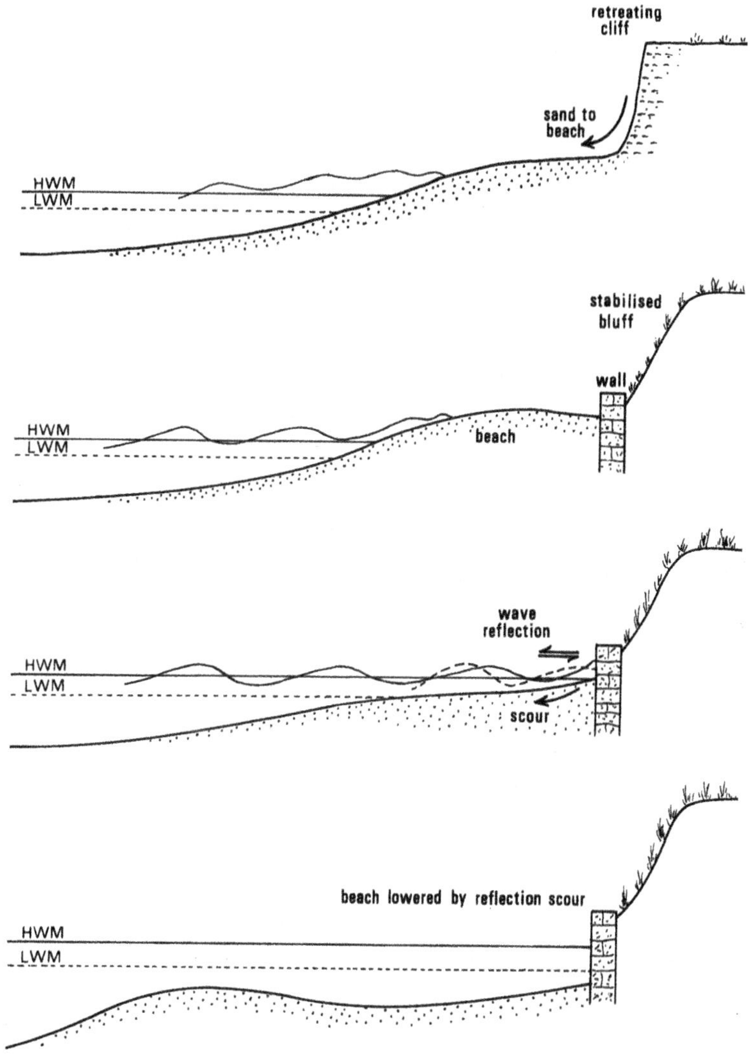

Fig. 2.15 Sequence where a beach fed with sediment from a receding cliff which is stabilised by building a basal sea wall. Waves reflected from the sea wall then withdraw sediment from the beach, which is lowered by reflection scour. © Geostudies

and gun emplacements, and massive sea walls were then constructed. These resulted in reflection scour, eroding the remaining beach, which is now completely submerged at high tide in St Ouen's Bay, Jersey (Fig. 2.18).

Calcareous beaches on the coast of Cornwall have been depleted by the extraction of shell sand and gravel for agricultural use as lime on farmland. Such extraction has traditionally been on a small scale (50–100 t/year) from several beaches, notably at Bude, where Summerleaze Beach has been depleted. Again,

Fig. 2.16 Beach at Surfers Paradise, Queensland, lowered by large waves reflected from a boulder wall during a tropical cyclone. © Geostudies

Fig. 2.17 Extraction of *sand* and *gravel* from the beach at Klim in Denmark. © Geostudies

Fig. 2.18 The sea wall at St Ouen's Bay, Jersey, with the beach lowered by reflection scour. © Geostudies

the artificial lowering of the shore profile has led to larger waves breaking on the shore, and increased beach erosion.

Erosion has developed on intensively used beaches at seaside resorts, which gradually lose sand as it is removed by visitors, adhering to their skin, clothes or towels, or trapped in their shoes. The losses are small, but cumulative, and no one brings sand to the beach. Pebbles and shells are also collected and carried away as souvenirs by beach visitors.

References

Bird ECF (1996) Beach management. Wiley, Chichester

Bird ECF (2011) Changes on the coastline of Port Phillip Bay. Victorian Government Department of Sustainability and Environment

Bruun P (1995) The development of downdrift erosion. J Coast Res 11:1242–1257

Bryant EA (1985) Rainfall and beach erosion relationships, Stanwell Park, Australia, 1895–1980: worldwide implications for coastal erosion. Zeitschrift für Geomorphologie, Supplementband 57:51–65

Cardno (2011) The great sands and adjacent coast and beaches. Report No. RM2289_LJ5518. Port of Melbourne Corporation

Christiansen C, Christoffersen H, Dalsgaard J, Nornberg P (1981) Coastal and nearshore changes correlated with die-back in eel grass (Zostera marina). Sediment Geol 28:163–173

Douglas BC, Kearney MS, Leatherman SP (2000) Sea level rise: history and consequences. International geophysics series, vol 75. Academic Press, London
El-Asmar HM, White K (2002) Changes in coastal sediment transport processes due to construction of new Damietta Harbour, Nile Delta, Egypt. Coast Eng 46:127–138
Elko NA, Davis RA Jr (2006) Morphologic evolution of similar barrier islands with different coastal management. J Coast Res (Special Issue 39):127–131
Elko NA, Holman RA, Gelfenbaum G (2005) Quantifying the rapid evolution of a nourishment project with video imagery. J Coast Res 21:633–645
Finkl CW, Esteves LS (1998) The state of our shores: a critical evaluation of the distribution, extension, and characterization of beach erosion and protection in Florida. Florida Shore and Beach Association, Tallahassee. FL, pp 302–318
Finkl CW, Walker HJ (2004) Beach nourishment. In: Schwartz M (ed) The encyclopedia of coastal science. Kluwer, Dordrecht, The Netherlands, pp 147–161
Grant US (1984) Influence of the water table on beach aggradation and degradation. J Mar Res 7:655–660
Loannidis D, Karambas TV (2007) Soft shore protection methods: beach drain system. In: 10th international conference on environmental science and technology, Kos Island, Greece, pp A-528–535
Lotfy MF, Frihy OE (1993) Sediment balance in the nearshore zone of the Nile delta coast, Egypt. J Coast Res 9:654–662
Orme AR (1985) California. In: Bird ECF, Schwartz ML (eds) The World's coastline. Van Nostrand Reinhold, New York, pp 27–36
Orviku K, Jaagus J, Kont A, Ratas U, Rivis R (2003) Increasing activity of coastal processes associated with climate change in Estonia. J Coast Res 19:364–375
Papps D, Priestley S (2005) Design and performance of replenishment projects on Auckland's eastern beaches. In: Proceedings of the 2005 coasts and ports conference, Australia, pp 125–130
Pezzuto PR, Resgalla JRC, Abreu JGN, Menezes JT (2006) Environmental impacts of nourishment of Balneario Camboriu Beach, SC, Brazil. J Coast Res (Special Issue 39):863–868
Ramana Murthy MV, Mani JS, Subramanian BR (2008) Evolution and performance of beach fill at Ennore Seaport, southeast coast of India. J Coast Res 24:232–243
Roberts TM, Wang P (2012) Four year performance and associated controlling factors of several beach nourishment projects along three adjacent barrier islands, west central Florida, USA. Coast Eng 70:21–39
Sestini G (1992) Implications of climate change for the Po delta and Venice lagoon. In: Jeftic L, Milliman JE, Sestini G (eds) Climate changes and the mediterranean. Arnold, London, pp 428–494
Temme B, Klein AHF, Carvalho JLB, Diehl FL (1997) Morphological behavior of beach of Balneario Camboriu: preliminary results. Notas Technicas da FACIMAR 1:49–65
Turner IL, Leatherman SP (1997) Beach dewatering as a soft engineering solution to coastal erosion—a history and critical review. J Coast Res 13:1050–1063
Wang P, Roberts TM, Dabees M, Horwitz MH (2011) Beach changes associated with active 2009–2010 El-Nino winter along the west-central Florida barrier islands. In Wang P, Rosati JD, Roberts TM (eds) Proceedings of Coastal sediments 2011. World Scientific, Singapore, pp 1229–1242

Chapter 3
Responses to Beach Erosion

Abstract Prior to considering beach renourishment, it is important to understand the potential alternative means of managing beach erosion, and consider beach renourishment in the context of one of many responses. Structural methods are briefly outlined before discussion on non-structural and adaptive methods.

As well as being necessary to know the causes of beach erosion prior to considering beach renourishment, it is also important to understand the potential alternative means of managing beach erosion, and consider beach renourishment in the context of one of many responses to beach erosion. Responses to beach erosion have historically been through structural engineering methods, but in the past century, and particularly the past few decades, adaptive and non-structural methods have become far more widespread. Each of the two general responses to beach erosion are briefly outlined.

Attempts to counter beach erosion have often been stimulated by a particular storm. For example, in Port Phillip Bay, Australia, a storm surge in 1934 caused extensive coastal damage, and the response of the local government was to set up a Foreshore Erosion Board (Mackenzie 1939), which surveyed the damage and decided which sectors required coastal defence works. In due course, this led to a beach renourishment programme (Sect. 4.6, p. 90).

3.1 Structural Engineering

A widespread response to beach erosion has been to build solid structures designed to protect and maintain existing beaches, or prevent further recession of a coastline. The two aims have often proved incompatible: retreat of a coastline can be halted by building structures such as sea walls or boulder ramparts along the eroding land margin, but this often leads to the depletion of beaches in front

E. Bird and N. Lewis, *Beach Renourishment*, SpringerBriefs in Earth Sciences,
DOI 10.1007/978-3-319-09728-2_3

of the structure by wave reflection (Sect. 2.14, p. 24), or by reducing a previous supply of sediment from the eroding coastline. Long-term monitoring of coastal change around structures built in response to beach erosion has frequently been shown to produce adverse environmental effects adjacent to, and on shores further away (Hamm et al. 2002; Bruun 1995). Exceptions to this have been discussed and illustrated by Basco et al. (1997) and Rakha and Kamphuis (1997).

Primitive sea walls are often banks of earth or other locally available material, particularly where it can be extracted from an adjacent excavation, often a parallel ditch. Subsequent reinforcement may be necessary in response to damage of such banks by storm waves, and may be followed by stages in the evolution of larger, more competent and costly solid structures.

Sea walls and boulder ramparts are generally introduced to prevent wave attack on an eroding coast, usually a receding cliff, an undermined and slumping bluff, or a truncated dune, in each case fronted by a beach that has not been sufficiently high and wide to prevent waves reaching the back of the shore.

Modern sea walls are usually large stone or concrete walls designed to withstand the force of the breaking waves, which are reflected seaward, and may scour away the beach. An alternative is to build boulder ramparts, also known as revetments or riprap, some of which are irregular heaps of rocky debris, others more carefully fitted blocks arranged on a seaward slope, usually on a mattress of sand or gravel. Tetrapods, made of reinforced concrete, are shaped to interlock and remain in position on the shore during phases of strong wave action. Boulder ramparts and tetrapods are less reflective than solid sea walls, and the expectation is that waves will break into crevices, producing swash and backwash that do not cause erosion on their seaward sides. In some places, rocky debris has been dumped to protect an earlier sea wall from undermining and disintegration by wave attack, or in the hope that a boulder apron would prove less reflective than a solid sea wall, and permit some recovery of a lost or diminished beach.

Sea walls and boulder ramparts are subject to damage by the impact of waves during storms (Fig. 3.1), and by scour due to the hurling of sand and gravel against

Fig. 3.1 Storm-damaged sea wall at Black Rock, Victoria, Australia. © Geostudies

them by large waves. Sea walls also decay by physical, chemical and biological weathering, a process that can be rapid on certain sandstones and limestones.

Construction of sea walls and similar structures on a particular sector of coast to protect a building or a seaside resort is usually followed by continuing recession on adjacent sectors, so that in due course the protected area becomes a promontory. This has happened at the seaside resort of Mundesley, on the East Anglian coast, which now protrudes between retreating cliffs of soft glacial drift, and at Bray on the dune coast of north-eastern France. Eventually the flanks of such promontories have to be stabilised artificially, and in due course the protected area could become an island.

Another response to beach erosion has been to introduce structures designed to retain a beach that protects a coastline from strong wave action. A breakwater built out from the coastline can intercept longshore drift in order to form a higher and wider beach sector updrift for some distance along the shore. The prograded sector becomes triangular and extends towards the outer end of the breakwater.

Multiple groynes (groyne fields), built of timber, masonry, sheet metal, boulders or concrete, have been inserted on some coasts, especially at seaside resorts such as Eastbourne in Sussex (Fig. 3.2), with the aim of retaining longshore drift and so protecting the coastline. Beach sand and gravel are intercepted in the intervening compartments, and accumulate until sediment spills over or round each groyne. Sometimes a larger terminal groyne is built at the downdrift end (Fig. 3.3).

As sediment drifting along the shore is trapped by the groynes the supply to downdrift beaches is reduced, and erosion is thus transferred along the coast. There is then a temptation to extend the groyne field: there are sectors of the coastline of England and Wales that now have multiple groynes for several miles.

Groynes have been successful in retaining beaches on some coasts, particularly where wave energy is generally low (Fig. 3.4), but storm waves may break in such a way as to withdraw sand or shingle seaward from beach compartments to the nearshore sea floor. Some of the withdrawn beach sediment may be returned during subsequent periods of calmer weather, but it is possible that longshore drift will carry it away along the nearshore zone, leaving the beach compartment between the groynes depleted (Fig. 3.5).

Fig. 3.2 Groynes defining beach compartments at Eastbourne on the Sussex coast, England. © Geostudies

Fig. 3.3 A terminal groyne retains the eastward-drifting beach at Hengistbury Head in Dorset. © Geostudies

Fig. 3.4 Groynes containing sandy beach compartments on the south coast of Botany Bay, New South Wales. © Geostudies

On some coasts breakwaters have been built at intervals along the shore to form artificial headlands between beach compartments. On the SE coast of Singapore a series of such headland breakwaters was built in the expectation that they would allow the dominant SW waves to shape a stable beach configuration within each

Fig. 3.5 Groynes on the Italian coast at Ravenna have failed to retain beach sediment, which has been withdrawn seaward and lost alongshore. © Geostudies

Fig. 3.6 Headland breakwaters separating beach compartments fronting reclaimed land on the south-east coast of Singapore in the 1970s. South-westerly wave action formed asymmetrical beaches between the breakwaters, but these did not remain stable, erosion continuing on the asymmetrical alignments. © Geostudies

compartment (Fig. 3.6). Waves arriving at an angle to the coastline would shape each beach into an asymmetrical 'half-heart' or zeta-curve configuration, which would be relatively stable, with minimal losses alongshore. Unfortunately, this procedure failed because erosion continued on the asymmetrical beaches between the groynes.

Another approach to beach protection has been to build nearshore or offshore breakwaters, detached structures parallel to the coastline, designed to interrupt and reduce wave action and induce beach accretion by waves refracted round the ends of the breakwaters, and also to shelter the accreted beach from erosive waves. At Borth on the Ceredigion coastline of Wales, construction of two offshore breakwaters and a submerged rock reef was completed in 2013 (Fig. 3.7). The structures accompanied beach renourishment. Incident wave patterns were modified so that the adjacent beach was protected from strong wave action, and widened in the lee of the breakwaters to form a cuspate spit. The standard of flood protection of the adjacent village was thus increased.

Offshore breakwaters are most effective on tideless shores, as around the Mediterranean Sea, where they do not have to face the problems of a regularly rising and falling sea. They are generally built parallel to the coastline, between 50 and

Fig. 3.7 Completed coastal protection works at Borth, United Kingdom. Beach renourishment supplemented a submerged rock berm and two breakwaters. © Royal HaskoningDHV

100 m long, and 50–200 m offshore, by stacking large boulders, usually of limestone or granite. Much of the seaside resort coast between Venice and Rimini in NE Italy has beaches lined by offshore breakwaters built of limestone blocks brought from the hinterland (Fig. 3.8). Waves break against them, and are diffracted through the intervening spaces before they reach the beach. This diffraction can lead to the shaping of cusps and shallow zones in the lee of each offshore breakwater, and if there is a sufficient sand supply these may grow into linking tombolos (Fig. 3.9).

Fig. 3.8 Chain of boulder breakwaters protecting the beach on the coast at Rimini, Italy. © Geostudies

Fig. 3.9 Tombolo formed in the lee of a breakwater on the coast of Singapore. © Geostudies

It may be possible to use mobile, floating breakwaters, which can be anchored in the nearshore zone as a means of inducing beach accretion, then towed away to allow the accreted beach to be washed and nearshore water cleaned by wave action. Various attempts have been made to construct floating breakwaters of rubber tyres, oil drums or timber, intended to reduce wave action and so diminish erosion or promote accretion on the beach, but these rarely survive the next storm.

Demands for the halting of cliff recession and beach erosion have thus peppered the world's coastline with an array of artificial structures of various kinds, some of which have been successful, others of little value: many have not been

Fig. 3.10 After a sea wall and groynes failed to retain a beach at Litorale di Pallestrina on the NE coast of Italy limestone boulders were dumped on the shore. © Geostudies

maintained, and are derelict. On the Lido di Pellestrina, near Venice, attempts were made to retain a wasting beach after a sea wall was constructed, then elaborated and armoured with boulders. Groynes were built, then a chain of offshore breakwaters, but as these defences multiplied the original sandy beach washed away, and has been replaced by dumped boulders (Fig. 3.10).

3.2 Non-structural Methods

Over the last few decades alternative means of constructing sea walls, boulder ramparts, breakwaters and groynes have become increasingly popular. Often termed 'soft engineering', these are intended to work with the existing processes, but despite this concept, they are not necessarily 'natural' nor always immune from environmental impacts, but offer an alternative to structural methods. The most widely applied method of 'soft engineering' is beach renourishment. In addition to beach renourishment, other adaptive methods on the beach include beach shaping and scraping, and use of covers to retain sediment.

Fig. 3.11 Large cobbles added to the beach at Amroth. Pembrokeshire, to increase the stability of the beach and increase the protection from flooding of hinterland properties. © Greg Guthrie of Royal HaskoningDHV

Dynamic revetments have also been used as an alternative to stone revetments or seawalls (Komar and Allen 2010; Ahrens 1990). Also termed "cobble berms" or "rubble beaches" they involve the construction of a cobble beach at the shore, in front of land or property to be protected. These structures are effective in defending the coastline as the sloping, porous cobble beach disrupts and dissipates wave energy by adjusting its morphology in response to the prevailing wave conditions. Dynamic revetments are therefore a variation of beach renourishment. An example of the introduction of a dynamic revetment was at Amroth in Pembrokeshire, United Kingdom where in the 1990s large cobbles were added to a shingle beach to give it additional stability (Fig. 3.11).

Other 'soft' methods of managing erosion include efforts to manage coastal dunes. As mentioned previously (Sect. 2.4, p. 14) dunes can supply a beach with sediment but also form a protective feature and therefore form an integral part of the beach system. Dune creation and restoration using vegetation and artificial methods to stabilise the dune system have been employed (Fig. 3.12). Dunes have been created and built up using clay cores (Wamsley et al. 2011), sand-filled bags (Komar and Allen 2010) and dredge material (Matias et al. 2005), all with varying levels of success. Other more primitive methods can be employed, for example in

Fig. 3.12 Planting of dune vegetation on Mona Vale Beach, Australia in 2013 in an attempt to halt erosion by restoring and stabilising the dune system. The stabilised dune is intended to protect coastal properties, avoiding the need for structural engineering or other 'soft' measures such as beach renourishment. © Nick Lewis

the United Kingdom unwanted Christmas trees have been dumped on the dunes as a means of trapping wind-blown sediment.

In addition to locally applied ‘soft’ measures of erosion management, more strategic approaches can be applied over greater geographical and temporal scales.

Managed retreat (or realignment) is becoming a greater consideration within in coastal management. Retreat involves major changes to land use and relocation of homes and infrastructure under threat (Few et al. 2007). Currently, managed retreat is applied mainly to estuarine environments and motivated by habitat creation, reducing the costs of flood protection and mitigating environmental impacts of coastal development. Managed realignment is being widely advocated in the United Kingdom for the alleviation of flood risk (Cooper 2003), and has been used in England and Germany (Rupp-Armstrong and Nicholls 2007). Methods include artificially created breaches in previously constructed embankments (Symonds and Collins 2007) and managed retreat of a previously developed or defended coastline (Cooper 2003). In Northern France and Belgium breaches have been made through dunes behind beaches so that the sea can enter through the barrier, reducing local erosion (Charlier et al. 2005). Managed retreat requires setting back coastal defences inland, and can have significant impacts on natural processes (French 1997, 2008).

In addition to ‘soft engineering’ more strategic coastal management, such as land use planning controls, can be employed to restrict future development in the coastal zone. For example in Australia setback lines of a selected distance for development projects are incorporated in planning policy (Walsh et al. 2004). Coastal construction setback lines, which limit new development in areas of high hazard, are used in many coastal programmes in the United States (NOAA 2000). By 2005 there were 466,620 ha (1,152,551 acres) of barrier islands ‘protected’ by United States law, but the Acts of Congress of 1982 and 1990 do not actually forbid development, they merely warn would-be builders that they cannot count on the Federal Government to provide insurance or reconstruction funds in case of damage due to storms, floods and erosion (Charlier et al. 2005).

References

Ahrens JP (1990) Dynamic revetments. In: Edge BL (ed) Coastal engineering 1990—proceedings 22nd coastal engineering international conference, Delft, The Netherlands, pp 1837–1850

Basco DR, Bellomo DA, Hazelton JM, Jones BN (1997) The influence of seawalls on subaerial beach volumes with receding shorelines. Coast Eng 30:203–233

Bruun P (1995) the development of downdrift erosion. J Coastal Res 11:1242–1257

Charlier RH, Chaineux MCP, Morcos S (2005) Panorama of the history of coastal protection. J Coastal Res 21:79–111

Cooper NJ (2003) The use of ‘managed retreat’ in coastal engineering. Eng Sustain 156:101–110

Few R, Brown K, Tompkins EL (2007) Climate change and coastal management decisions: insights from Christchurch Bay, United Kingdom. Coast Manag 35:255–270

French PW (1997) Coastal and estuarine management. Routledge, London

French JR (2008) Hydrodynamic modelling of estuarine flood defence realignment as an adaptive management response to sea-level rise. J Coastal Res 24:1–12

Hamm L, Capobianco M, Dette HH, Lechuga A, Spanoff R, Stive MJF (2002) A summary of European experience with shore nourishment. Coast Eng 47:237–264

Komar PD, Allen JC (2010) 'Design with nature'—strategies for shore protection—the construction of a cobble berm and artificial dune in an Oregon State Park. In: Proceedings of a state of the science workshop, US geological survey scientific investigations report [online] 2010-5254, pp 117–126

Mackenzie AD (1939) Coastal erosion in Victoria. Trans Inst Eng Aust 20:229–236

Matias A, Ferreira O, Mendes I, Dias JA, Vila-Concejo A (2005) Artificial construction of dunes in the south of Portugal. J Coastal Res 21:472–481

NOAA (National Oceanic and Atmospheric Administration) (2000) State, territory, and commonwealth beach nourishment programs: a national overview. Technical document 00-01, OCRM program policy series

Rakha KA, Kamphuis JW (1997) A morphology model for an eroding beach backed by a seawall. Coast Eng 30:53–75

Rupp-Armstrong S, Nicholls RJ (2007) Coastal and estuarine retreat: a comparison of the application of managed realignment in England and Germany. J Coastal Res 23:1418–1430

Symonds AM, Collins MB (2007) The development of artificially created breaches in an embankment as part of a managed realignment. J Coastal Res 50:130–134 (special issue)

Walsh KJE, Betts H, Church J, Pittock AB, McInnes KL, Jackett DR, McDougall TJ (2004) Using sea level rise projections for urban planning in Australia. J Coastal Res 20:586–598

Wamsley TV, Waters JP, King DB (2011) Performance of experimental low volume beach fill and clay core dune shore protection project. J Coast Res 59:202–210 (special issue)

Chapter 4
Beach Renourishment Principles

Abstract This chapter provides a discussion of fundamental beach renourishment principles, including the need for preliminary investigations, sources of sediment for beach renourishment, methods of beach renourishment, design considerations, techniques for monitoring changes, assessment of performance and modelling.

Experience from various beach renourishment projects and their design and implementation is here used for discussion of the following topics:

4.1. The need for preliminary investigations
4.2. Sources of sediment for beach renourishment
4.3. Methods of beach renourishment
4.4. Design considerations
4.5. Monitoring changes after beach renourishment
4.6. Assessment of beach performance
4.7. Modelling of beach renourishment
4.8. Beach renourishment for coast protection

4.1 The Need for Preliminary Investigations

It is necessary to decide where and why beach renourishment is necessary, where and how sediment should be delivered to the shore, and what difficulties are likely to be encountered during and after renourishment. A preliminary investigation should include:

- The dimensions and morphology (with transverse profiles) of the beach and nearshore (including the breaker zone).
- The relationship of the beach to nearby cliffs, bluffs, reefs, river mouths, tidal inlets and drains.

E. Bird and N. Lewis, *Beach Renourishment*, SpringerBriefs in Earth Sciences,
DOI 10.1007/978-3-319-09728-2_4

- Grain size (modal, range) and shape (rounded, angular): proportions of sand, granules, pebbles, cobbles, shells, rock fragments, alien particles (brick, glass, earthenware).
- Dominant minerals (quartz, feldspar, carbonates, olivine).
- Evidence of source(s) of beach sediment (continuing or relict).
- Wave regime, storm effects, seepage, runoff.
- Tide ranges (neap, spring).
- Evidence of longshore drift (dominant direction or seasonally alternating).
- Evidence of offshore-onshore movements of beach sediment.
- Presence of microcliffs, berms and washovers.
- Evidence of history (maps, remote sensing, previous surveys).
- Evidence of rates or erosion or accretion over a specified time
- Indications of cause(s) of beach erosion.

An example of an unsatisfactory beach renourishment project, apparently because of inadequate preliminary investigation, can be quoted from Half Moon Bay, on the NE coast of Port Phillip Bay, Australia. This bay extends between headlands at Red Bluff, to the north, and Black Rock Point, to the south, and contains an arcuate sandy beach about 450 m long and up to 25 m wide at mean high tide. It is fronted by shallow water with sand bars that diminish incident wave action, and is still receiving sand eroded from gullies in soft Red Bluff Sand in the cliff at Red Bluff. For this reason it had been the most stable of Melbourne's bayside beaches, apart from a limited response to alternating seasonal longshore drift: during summer southerly wave action drifts sand northward, widening the beach at the northern end and narrowing it to the south, while in winter westerly wave action reverses this drift (Sect. 4.3.2, p. 64). The alternations have been balanced, so that there have not been net gains or losses from this beach compartment in recent decades.

The cliff at Red Bluff has a basal outcrop of hard Black Rock Sandstone, which is exposed to occasional storm waves from Port Phillip Bay, but has changed very little in the past half-century. This Black Rock Sandstone is overlain by 25 m of

Fig. 4.1 Heaps of sand dumped on the beach at Half Moon Bay, Melbourne, Australia, prior to being spread on the renourished beach © Geostudies

softer Red Bluff Sand, dissected into gullies by runoff and seepage. It was suggested that the erosion of Red Bluff could be controlled by renourishing the sandy beach in Half Moon Bay in order to widen it about 10 m and allow it to extend round the base of the cliff as a protection against storm wave erosion (Fig. 4.1). Erosion on the cliff was largely due to gulleying and seepage, which was also delivering sand to the beach. The idea that renourished beach sand in Half Moon Bay would drift on to the coastline salient beneath Red Bluff and remain there was doubtful because of the problem of maintaining a beach on a coastline salient, where the normal condition is drift divergence in response to incident waves arriving parallel to the coastline (Fig. 4.2a). Elsewhere, a beach may persist on a coastal salient (promontory or headland) if obliquely-arriving waves maintain longshore drift past it (Fig. 4.2b).

In October 2011 contractors began dumping lorry-loads of sand brought from an inland quarry on the beach at Half Moon Bay. There were protests by local people and the beach nourishment project was halted with about 60 heaps of sand dumped on the beach (Fig. 4.1). The protests were about the renourishment of a

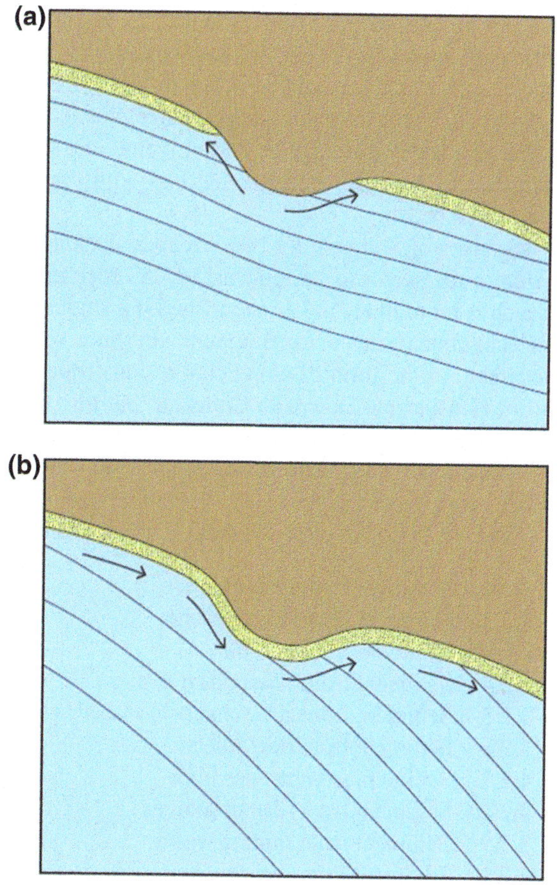

Fig. 4.2 **a** Where waves arrive parallel to the coastline they produce a divergence of drift (*arrowed*) on a coastal salient (headland), so that a beach is not maintained there. **b** Where waves arrive at an angle to the coastline they produce longshore drift, so that a beach may persist round a coastal salient, maintained by sediment in transit

beach that had been stable, in contrast with other nearby beaches that did need renourishment and about the colour of the dumped sand (yellow with an iron oxide stain: sand from the same quarry dumped on another Port Phillip Bay beach at Rye had quickly lost this stain by wave wash and rainfall) and its coarser texture (which is actually good management practice, see below). The suggestion that beach nourishment in Half Moon Bay could reduce erosion on Red Bluff was based on incorrect assumptions about the processes at work on the cliff.

In November 2011 the sand heaps were retrieved by lorries and taken round to another beach at Altona, on the far side of Port Phillip Bay, where renourishment was necessary.

4.2 Sources of Sediment for Beach Renourishment

Almost any kind of durable sediment of suitable grain size can be used for beach renourishment, providing it does not contain pollutants or hazardous items such as broken glass fragments or jagged metal. Sediment is obtained from sites known to engineers as 'borrow areas', a curious term which is only really appropriate in situations where the sand and gravel are expected to drift back where they came from. It would be better to refer to them as 'source areas'. Ideally the sediment should have similar grain size characteristics to the natural beach sediment (James 1974; USACE 2002); if not, allowance must be made for the rapid removal of the finer constituents and adjustments in profile as waves sort the emplaced beach.

Sources of sediment for beach renourishment should be sought as close as possible to the sector to be replenished, in order to minimise transportation costs. A beach is more likely to be renourished if a source of suitable material exists within an economic distance: for Bournemouth (Sect. 4.2.7, p. 49) the ideal beach fill was available in the form of china clay residue (quartz and feldspar sand and gravel) from Hensbarrow Downs in Cornwall, but this was over 200 km away, too far to be economically worthwhile, and it was necessary to use sand and gravel dredged from the nearby sea floor (Willmington 1983) and later from within adjacent Poole Harbour (Cooper 1998).

The various sources of sediment for beach renourishment are as follows:

- 4.2.1. Sediment from land quarries
- 4.2.2. Sediment from other beaches
- 4.2.3. Sediment from harbours
- 4.2.4. Sediment from coastal lagoons
- 4.2.5. Sediment from river channels and alluvial plains
- 4.2.6. Sediment from tidal inlets
- 4.2.7. Sediment from the sea floor
- 4.2.8. Sediment from distant sources
- 4.2.9. Sediment from mining waste
- 4.2.10. Sediment from recycled materials

4.2.1 Sediment from Land Quarries

Sand and gravel have been obtained from quarries and trucked down to the coast to renourish beaches in Monterey Bay, California and Ediz Hook on the Washington coast of the United States (Sect. 4.3.1, p. 61), Michigan City on the Great Lakes (Sect. 4.5, p. 87), Sidmouth on the south coast of England and Redcliffe in Queensland, Australia. Black Rock beach, on the shores of Port Phillip Bay, Australia, was partly renourished in this way (Fig. 4.3). Erosion on Bramston Beach in Queensland, was offset by bringing in 6,000 m^3 of coarse sand quarried from old beach ridges (with the advantage that it had previously been a beach sediment) a short distance inland and delivered to the shore by lorries. The use of land-based sources of sediment for beach renourishment may lead to problems with lorry traffic. The cost of using such sediment for beach renourishment may also be too high where there are competing demands for sand and gravel extracted from quarries for other purposes such as road making or concrete aggregate.

In 2008 beach renourishment at Camber Sands in East Sussex, SE England, used sediment from a nearby gravel pit (CIRIA 2010), chosen as it provided a number of advantages over alternative sources: the sediment was of similar calibre to the natural beach sand, traffic movement was reduced, and the sediment was cheaper than that obtained from other more distant sources.

4.2.2 Sediment from Other Beaches

Sand or gravel can be brought alongshore from other beaches, particularly those that have been widened by progradation, to renourish a depleted beach. At Aberystwyth in Wales the resort beach was augmented in 1963 by shingle brought from a beach to the north (So 1974). At Hvidesande, in Denmark, southward

Fig. 4.3 Sand dumped from lorries on the shore at Black Rock, Victoria, Australia. © Geostudies

drifting sand accumulated alongside breakwaters built in 1910 to stabilise the entrance to a coastal lagoon known as Ringkøbing Fjord and ensure ship access to the port of Ringkøbing on the shores of that lagoon. Sand extracted from the widened beach north of the breakwaters was trucked southward to renourish the downdrift beach, which had been cut back up to 45 m as the result of the drift interception (Møller 1990). This transfer of sand from an accreted beach updrift to an eroded beach downdrift is an example of bypassing (Sect. 4.3.3, p. 65), while sediment moved from an accreted beach downdrift back to a depleted beach updrift is termed recycling (Sect. 4.3.2, p. 66).

4.2.3 Sediment from Harbours

A major source of sediment for beach renourishment has been sand and gravel dredged from harbours and port approaches. The first such project was in 1922, when 1.3 million m^3 of sand dredged from New York Harbour was deposited on a 1 km stretch of beach on nearby Coney Island (Hall 1952; Dornhelm 1995).

Following construction activity at Ennore Port on India's east coast, 3.5 million m^3 of dredged sediment was transported through a pipeline to a beach immediately north (Ramana Murthy 2008). The sediment provided defence against erosion, apparently as a result of the port construction, as well as a cost-effective use of dredge material. Out of the total quantity of dredged sediment, 700,000 m^3 was placed on the upper beach to raise the berm crest, with the remainder spread across the nearshore zone to increase the beach width by 500 m.

During the winter of 2005–2006, 1.1 million m^3 of sand dredged from the entrance channel of Poole Harbour were used to renourish the nearby beaches of Swanage, Poole and Bournemouth.

Sand dredged to maintain a navigable boat channel at Barnegat Inlet, New Jersey was used in 1979 to renourish an eroded ocean beach on nearby Long Beach Island (Psuty 1984), and 2 million m^3 of sand obtained from a similar boat channel was used to restore the beach on Sandy Hook, New Jersey, where a spit had been breached by erosion in 1982.

Sand dredged from the harbour at Rio de Janeiro in the 1940s was used to renourish Copacabana Beach, which was meagre until it was thus enlarged to provide a wide shore recreational area above normal high tide level (Vera Cruz 1972). At Robe, in South Australia, sand dredged to maintain the port approach was dumped to restore an adjacent beach.

Several beach renourishment projects in New Zealand have used sediment available as a by-product of port dredging or marina projects (Healy et al. 1990). Dredging at the ports of Napier, New Plymouth and Tauranga Harbour produced large quantities of sand and gravel, which have been used to renourish beaches downdrift.

4.2.4 Sediment from Coastal Lagoons

On barrier island coasts the backing lagoons may provide a suitable source of sediment for outer beach renourishment. About 450,000 m^3 of relatively coarse sand dredged from a lagoon floor behind a coastal barrier was used to renourish an eroded beach north-east of Atlantic City, New Jersey, in 1963 (Sect. 4.6, p. 87). Sediment was dredged from backing lagoons to renourish beaches on the outer shores of Hel spit in Poland, while at Beachport in South Australia, sand dredged from a lagoon entrance has been used to renourish the town beach. In East Australia, sand dredged from the entrance of Narrabeen Lagoon has been placed on the adjacent beach. Dredging at the entrance to the lagoon was primarily a flood mitigation measure for properties located in the Narrabeen Lagoon floodplain, with dredging maintaining the capacity of the lagoon entrance to discharge floods.

Erosion of the beach at Lido di Jesolo, Italy, on one of the barrier islands fronting the Venice Lagoon, was countered by building sea walls, groynes and offshore breakwaters, but when these failed to retain the beach it was replenished by dumping sand dredged from the floor of the lagoon (Zunica 1990).

At Odessa on the Ukrainian Black Sea coast, some 3 million m^3 of sediment excavated for the building of Port Yuzhniy, in a nearby coastal lagoon, 30 % of which was limestone gravel and the rest sand and clay was used to replenish eroded beaches (Shuisky 1994).

4.2.5 Sediment from River Channels and Alluvial Plains

Sediment removed from rivers has been used for beach renourishment. Sediment from a shoal at the mouth of the Camboriu River, southern Brazil was used to supply 50,000 m^3 of sediment to renourish the adjacent Balneario Camboriu Beach (Pezzuto et al. 2006). This beach renourishment was used to mitigate the loss of beach material caused by a number of strong erosional events, combined with urbanisation of the coastal plain.

Sediment dredged from rivers draining the Caucasus Mountains was used to renourish beaches on the Georgian Black Sea coast, which were originally supplied naturally with sand and gravel from these rivers. By 1970 the natural beaches had been depleted by extraction of sand and gravel for building purposes, damming of the rivers, and the building of harbour breakwaters. Between 1981 and 1987 six beaches on 47.5 km of coastline between Gagra and Batumi were renourished with 9.2 million m^3 of sand and gravel excavated from river channels and alluvial plains (Zenkovich and Schwartz 1987). On the Turkish Black Sea coast several beaches have been augmented by the dumping of sand and gravel obtained from steep mountainous rivers, mainly to protect the coastal highway.

On the shores of the Barron delta north of Cairns, Australia, beaches had been naturally maintained with sand washed out from distributaries of the Barron River, but erosion began on Holloways Beach when the mouth of a distributary that had been supplying sand was diverted northward to Richter Creek during a 1938 flood. Deprived of a sand supply, Holloways Beach began to erode as sand drifted away northward. In 1992 a 600 m sector of this beach was renourished with 83,000 m^3 of sand dredged from the lower reaches of Richter Creek and delivered to Holloways Beach by lorries, essentially an artificial revival of the fluvial sand supply that originally built this beach. Maintenance of such deltaic beaches could also be achieved by relocating the mouth of a river distributary so that it fed sand to the updrift end of a resort beach.

Sand from the Mississippi River has been used to restore beaches and dunes on Scofield Island, a rapidly-deteriorating barrier island in Plaquemines Parish, Louisiana. As part of the project planning, Poff et al. (2011) evaluated the feasibility of mining and transporting sand from the Mississippi River. Riverine sand was recommended as an alternative to offshore sources.

4.2.6 *Sediment from Tidal Inlets*

On barrier island coasts there are often shoal deposits at tidal inlets, formed seaward by ebb tides and lagoon ward by flood tides. It is necessary to consider the possible impacts of extracting sediment from such shoals, such as the effects on wave and current regimes in the inlet, and the possibility that changing wave patterns will have adverse effects on the adjacent coastline. At Captiva Island in Florida resort beaches were renourished with sand dredged from ebb-tide shoals off the nearby inlet at Red Fish Pass. This extraction proved beneficial because it modified the wave refraction pattern in such a way as to stop sand drifting back from the renourished beach into the inlet. Sand was also available from flood-tide shoals just inside Red Fish Pass, but this was not used because of the risk that deepening this area would induce sediment inflow and result in erosion alongside the inlet (Walton and Dean 1976). At Palm Beach on Australia's Gold Coast, sediment has been dredged from inside the mouth of nearby Currumbin Creek and used to renourish the southern end of Palm Beach. Dredging has taken place at least once a year for the last three decades. Sediment is pumped onto the beach by pipes then a bulldozer is used to distribute the sand along the beach.

4.2.7 *Sediment from the Sea Floor*

Sand and gravel dredged from the sea floor has been widely used for beach renourishment in such places as Seaford in England, Ostend in Belgium (Sect. 4.3.1, p. 56), Port Dickson in Malaysia and Miami Beach in Florida (Sect. 4.3.8, p. 73/74). Nearshore shoals have provided sandy sediment for beaches on Nordeney in Germany (Sect. 4.4.2, p. 80), on the Adelaide coast in South Australia (Sect. 4.3.4,

p. 67), and Brest in France. There are sometimes alternative demands for sediment extracted from the sea floor (e.g. for aggregate or building purposes), but this is less likely where the sediment contains clay, chalk or shelly debris, which are not suitable for construction but can be used for beach fill.

In the United Kingdom, many of the beach recharge projects have used sediment from the seafloor of the continental shelf. Licensed aggregate dredge areas are leased by the Crown Estate to aggregate supply companies along with a permitted annual tonnage. It was estimated in 2009 that potentially workable resources of UK offshore sand and gravel exceeded 800 million m^3 (CIRIA 2010).

The beach at Bournemouth on the south coast of England was depleted after a concrete promenade was built to stabilise eroding cliffs, thereby halting the supply of sand and gravel to the beach (Fig. 2.3). This beach was renourished in 1974 with coarse sand and gravel dredged from the sea floor off the Isle of Wight and dumped in a zone 450 m off the Bournemouth shore. This was then delivered to the beach through a floating steel pipe by a pump mounted on a pontoon. Beach compartments between numerous groynes were renourished, using 830,000 m^3 of sand and gravel to form a beach 5 miles long. Longshore drift carried some of the beach material away to the east, but Bournemouth beach has been subsequently maintained by dumping more sand and shingle obtained from the sea floor on the shore at intervals. Subsequent renourishment of Bournemouth Beach in 2006 used sediment dredged from the entrance to Poole Harbour, as mentioned above.

The largest beach renourishment scheme undertaken in the United Kingdom is along the Lincolnshire coast between Mablethorpe and Skegness, and has been termed 'Lincshore'. Between 1994 and 1998 a total of 7.6 million m^3 of sand and gravel were dredged from offshore banks and placed on the coast (Duvivier 1998). A further 240,000 m^3 of similar sediment was placed at two locations in 1999 to replace losses following the first placement (Blott and Pye 2004). The renourishment was principally undertaken as part of a strategy to protect flood defences, but it was also considered that the method would improve the aesthetics and support tourism (Duvivier 1991). This is an ongoing project and is currently nearing the end of a 5 year (2010–2015) programme of beach renourishment. Since 2010 a further 1.5 million m^3 of sand has been placed on the beach at a number of locations along 24 km of coast, at a rate of around 11,000 m^3/day. Sand is delivered through hydraulic pumping followed by shaping by excavators and bulldozers.

Sand dredged from the sea floor was used in a similar way to restore Mentone Beach on the north-east coast of Port Phillip Bay, Australia (Sect. 4.3.1, p. 55), which had been depleted by reflection scour from a sea wall (Fig. 4.4). Sea floor sand immediately offshore was too fine and silty used to be used for this purpose, but between 1.4 and 2.0 km seaward, where the water was 7–10 m deep, there was a deposit of coarser sand up to a metre thick, containing marine shells (Guerin 1984), which was pumped on to the shore to renourish the beach (Fig. 4.5).

At Burleigh Heads in Queensland, Australia, 100,000 m^3 of sand dredged from the sea floor 1.5 km offshore (where the water is 18–25 m deep) in 1985 was brought in and dumped on sand bars in the nearshore zone. Within a year much of this sand had been washed on to the beach.

Fig. 4.4 The shore at Mentone, Port Phillip Bay, in 1976 when the beach had been depleted by reflection scour in front of a sea wall. This beach was then renourished by pumping in sand from the sea floor (see Sect. 4.3.1) © Geostudies

Fig. 4.5 Depleted beach at Mentone, Port Phillip Bay, Australia, after sea wall construction. © Geostudies

On some coasts, notably in Florida, it has been found that extraction of sand or gravel from the sea floor for beach renourishment causes subsequent erosion because of increased wave energy through deepened nearshore areas.

4.2.8 Sediment from Distant Sources

Where sufficient quantity or quality of sediment is not available within a reasonable distance from the beach, sediment may need to be sourced from a distance.

Coarse sand and gravel dredged from coralline shoals, reefs and cays and ferried in to the coast has been used for beach renourishment on a number of resorts on Caribbean islands, while sand and gravel dredged from coral lagoons was used to renourish eroding resort beaches on cays at Green Island, off Cairns in north-eastern Australia. In Cuba the beach at the seaside resort of Varadero, on the Hicacos Peninsula north-east of Havana, suffered severe damage on the winter of 1986–1987 when storm waves washed along the hotel fronts. In 1990 81,000 m^3 of coarse sand was obtained from Cayo Mono, an uninhabited coral cay 25 km offshore, brought in by barge, pumped onshore, then shaped by bulldozers to restore a recreational beach 1.4 km long (Schwartz et al. 1991). In Brazil, sand of a suitable size for renourishment of beaches was sought on the continental shelf SE of Rio de Janeiro (Oliveira and Muehe 2013). With the development of large dredging barges, sourcing of sediment from distal offshore areas has become more common.

In Malta, there are no significant offshore sources of sand available. Instead, sand for use in beach renourishment must be crushed from rock originating from overseas quarries as local limestone is too friable (Firman et al. 2011).

Sand was brought from a distance to improve Waikiki beach at Honolulu in Hawaii (Finkl and Walker 2004). The natural beach at Honolulu was originally rather meagre on a rocky, reef-fringed shore, until it was renourished in several stages beginning in the 1920s. There are conflicting reports on where the sand came from, but apparently some sand was shipped in from Manhattan Beach, in California, and there are anecdotes of importation from various other sources, including Australian beaches (Campbell and Moberley 1984). In the Canary Islands some tourist beaches at seaside resorts, such as San Andreas in Tenerife, have been artificially nourished with sand shipped from the Sahara Desert.

The shingle beach in the bay at Anne Port, on the east coast of Jersey, has a shingle beach that became depleted by reflection scour in front of a sea wall. In 1999 it was renourished with well-rounded pebbles imported from a quarry near Cork in southern Ireland, chosen because of their similarity to the preceding natural beach at Anne Port (Fig. 4.6).

Fig. 4.6 The shingle beach at Anne Port, Jersey, renourished with gravel shipped from Ireland. © Geostudies

4.2.9 Sediment from Mining Waste

Mining waste (sand and gravel) from tin mines in south Cornwall was piped through a tunnel cut in 1842 in a coastal ridge to form an artificial beach in Carlyon Bay (Figs. 4.7 and 4.8) (Everard 1962). The beach was eventually about 200 m wide, with a broad convex profile, but as mining declined in the 20th century the fluvial sediment supply diminished, and artificial nourishment of Carlyon Beach ceased. The beach remains as one of the earliest and most substantial artificially nourished beaches in Britain, but when the supply of mining waste halted the beach began to erode, developing a concave profile. Continuing erosion of this beach is likely to pose a problem for those intent on developing a seaside resort here.

An example of beach renourishment prompted by the need to dispose mining waste has been reported from Chañaral Bay in southern Chile (Paskoff and Petiot 1990). Between 1938 and 1975 tailings from a copper mine were dumped on the shores of Chañaral Bay, on the Atacama Desert coast, at the rate of more than 4 million m^3/year. They included silt and clay, which were swept offshore, but the sandy fraction was retained to prograde the beach by an average of 900 m over this period. After 1975 the dumping was transferred to Caleta Agua Hediona, the next

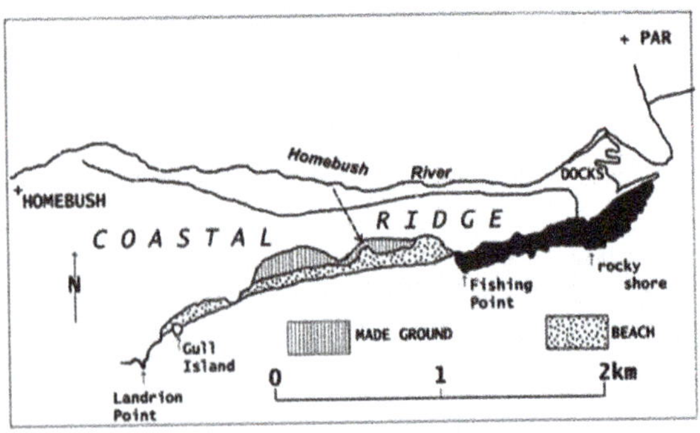

Fig. 4.7 The Carlyon Bay coast, showing the Homebush valley and the tunnel through which mining waste was diverted to the shore to prevent it from choking the port of Par, to the east. © Geostudies

Fig. 4.8 The artificial beach in Carlyon Bay, Cornwall, was supplied with sand and gravel waste from tin mining in the hinterland, delivered through a tunnel cut from the Homebush valley. © Geostudies

bay to the north, which had a rocky shore with stacks in front of cliffs. By 1985 this had acquired an artificial beach more than 5 km long and up to 300 m wide.

A major potential source of sediment suitable for beach renourishment in south-west England exists in the tip-heaps of quartz and feldspar sand and gravel in the china clay quarrying region of Hensbarrow Down, near St Austell in Cornwall. There is a similar, smaller area in south-western Dartmoor, which was used to replenish a small beach at Torpoint, near Plymouth.

Colliery waste was dumped on the shores of County Durham, England for more than a century, until the closure of the coal mines in recent years (Hydraulics Research Station 1970; Nunny 1978). Deposits of dumped waste have been reworked and drifted alongshore for several kilometres, augmenting the natural beaches of locally-derived calcareous sand (Fig. 4.9). The added coal and shale have been sorted into gravel and sand, and in some places coal fragments were collected from the shore, mainly for domestic use. Wide, convex beaches formed, but after dumping was halted, the beaches were cut back by wave action, and low cliffets or beach scarps were formed, fronted by a developing concave profile.

Fig. 4.9 The beach at Ness Point, near Seaham in Durham, was augmented by the dumping of gravelly waste from nearby coal mines. Dumping ceased when nearby coal mines were closed in 1993, and the beach is being re-shaped by wave action. A low clifflet (*arrowed*) marks scarping at the limit of wave re-working. © Geostudies

Waste from a steelworks has been used to renourish beaches at Port Lincoln and Port Augusta, in South Australia. Ships ballast was added to the beach in Oriental Bay, on the north-east shore of Wellington Harbour, New Zealand, to improve it for recreation.

4.2.10 Artificial Sediment

The possibility of using artificial sand for beach renourishment has been considered in Japan, where a laboratory has tried to produce foraminiferal sand ("star sand"). Alternatively, broken glass (cullet) can be ground into sand-sized fragments (Edge et al. 2002; Makowski and Rusenko 2007; Makowski et al. 2011). First proposed by Finkl (1996) for use on Florida's beaches, recycled glass cullet has been found to retain the same physical properties as natural silica sand and can be mechanically processed to match the grain size of the existing beach sediment. Recycled glass has also been used on beaches along Lake Hood in New Zealand, the Dutch Caribbean island of Curacao and Hawaii (Williams and Micallef 2009).

4.3 Methods of Beach Renourishment

Methods of renourishing a beach vary according to the configuration of the coast and the processes at work on it. Locations of renourishment within a beach also vary and include dune nourishment; nourishment of the subaerial beach (berm), profile nourishment (subaerial and submerged) and bar or shore face nourishment (submerged fill).

Direct placement of fill can be used on a coastal sector, particularly where longshore drift is weak, or can be controlled by the insertion of groynes. Renourishment should be at points or sectors from which it is expected that longshore drift will carry the sediment to where it is required. There is bypassing, where sediment is conveyed from a sector that has prograded alongside an obstacle such as a breakwater, a river mouth or a tidal inlet, along the coast to replenish a beach depleted downdrift. There is recycling (Sect. 4.3.4, p. 66), whereby beach losses due to longshore drift are made good by bringing back the sediment. There is nearshore renourishment based on the expectation of shoreward drift, and backpassing, which is analogous to longshore recycling in that it brings back sediment lost seaward from a beach. Offshore breakwaters have been used to renourish a beach by inducing accretion in their lee, and some beaches have been emplaced or reshaped by bulldozing. Each of these techniques will be considered and exemplified in the following sections.

- 4.3.1. Direct placement
- 4.3.2. Emplacement by longshore drift
- 4.3.3. Bypassing
- 4.3.4. Recycling

4.3.1 Direct Placement

Some beaches have been renourished by dumping truck-loads of sand on the shore, as at Morib in Malaysia (Fig. 4.10).

Mention has been made (Sect. 4.2.7, p. 49) of the beach at Bournemouth on the south coast of England, which was renourished by pumping sand in from a nearshore stockpile directly on to the depleted beach sector. A similar project was carried out at Mentone, near Melbourne, Australia, on a coast that is much like Bournemouth geologically. Receding cliffs in soft Tertiary sandstones were stabilised by a sea wall and promenade built in 1937–1939, and the sandy beach that had been supplied with sediment eroded from the cliffs then diminished. In 1977 the beach was renourished with about 160,000 m^3 of sand dredged from a rectangular zone 1,800 m long and 600 m wide the sea floor (Fig. 4.11) and pumped on to the shore (Fig. 4.12). The sediment supplied to the shore was bulldozed to form a beach terrace, initially about 32 m wide, built 2 m above low spring tide level. The sand was at first dark in colour because of a coating of silt, clay and organic matter, but rain quickly washed this away. There was a problem because the beach had been emplaced across several storm water outfall pipes, and the first heavy rain saw gullies washed out across the beach by outflow. This was remedied by extending the pipes seaward to the outer edge of the renourished beach.

Fig. 4.10 Renourishment of the beach at Morib, Malaysia, by dumping sand from lorries. © Geostudies

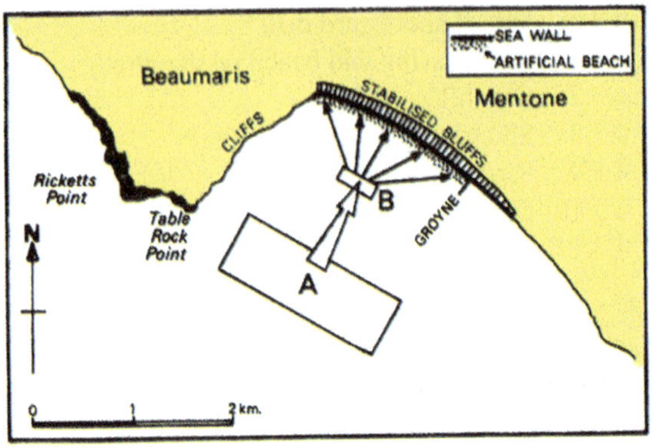

Fig. 4.11 The coast at Mentone, Port Phillip Bay, Australia, showing the area from which sea floor sand was dredged (*A*) and the zone where it was dumped (*B*) to be pumped on to the shore (see Fig. 4.12). © Geostudies

Fig. 4.12 Sand pumped on to the shore at Mentone, Port Phillip Bay, Australia, and deposited as a renourished beach. © Geostudies

There are limits to the distance over which pumping of sand is effective. On the Belgian coast 500,000 m^3 of sand dredged from the fishing port at Ostend was pumped on to the shore between Bredene and Klemskerke in 1978 to renourish a depleted beach up to a kilometre to the west, but beyond this it was necessary to pump from another source, coarse shelly sand from the offshore shoals at Stroombank and Kwintebank. Some 8.5 million m^3 of sand delivered from these sources permitted beach placement on 8 km of coast near Zeebrugge (Kerckaert et al. 1986).

More recently at Ostend, beach renourishment has been undertaken in 2013 using sediment dredged from offshore sand banks. A mix of water and sediment was pumped aerially directly from the dredging vessel onto the beach or nearshore zone. This process also known as 'rainbowing' due to the shape of the slurry as it passes through the air (Fig. 4.13) is widely used where sediment has been dredged from the sea floor or from estuaries and inlets along the coast.

As a sequel to the Bournemouth project, several other seaside resorts on the south coast of England have had their beaches renourished in front of esplanades. Some have used boulder armouring to reinforce the sea wall, and covered this with imported shingle to form a new beach.

Fig. 4.13 Beach renourishment taking place at Ostend, Belgium in 2013. Sand extracted from the seafloor was pumping directly from the dredge barge to renourish the beach and nearshore areas. © Afdeling Kust

At Seaford in East Sussex beach erosion began after the building of the large breakwater at Newhaven in 1845, which cut off the supply of shingle drifting in from the west. As Seaford beach was depleted, storm waves became increasingly destructive, and a sea wall was built to protect the esplanade, with numerous groynes inserted in the hope of retaining what was left of the shingle (Fig. 4.14), which was augmented artificially in 1963. Beach erosion continued despite successive elaborations of the structures. Eventually, in 1987, the shingle beach was renourished by the Southern Water Authority with 1.5 million m^3 gravel dredged from the sea floor off Littlehampton, to the west, dumped a kilometre offshore, then pumped on to the shore, where it was deposited over an armouring of large granite blocks imported from Galicia in Spain. The restored shingle beach was then shaped by bulldozers into a broad terrace with a seaward outer slope (Fig. 4.15), and a retaining groyne at the eastern end (Nicholls 1990).

A similar technique has been used to restore the depleted shingle beach at Sidmouth in Devon. Protected by sea walls built in the 19th century, this south-facing seaside resort had been steadily losing its beach, partly because the adjacent cliffs of soft sandstone have continued to recede, so that the esplanade stands slightly forward from the general coastline. In consequence, the seafront was more exposed to wave scour, which dispersed the shingle beach alongshore, mainly to the east. It was decided to reinforce the sea wall with a rampart of large stone

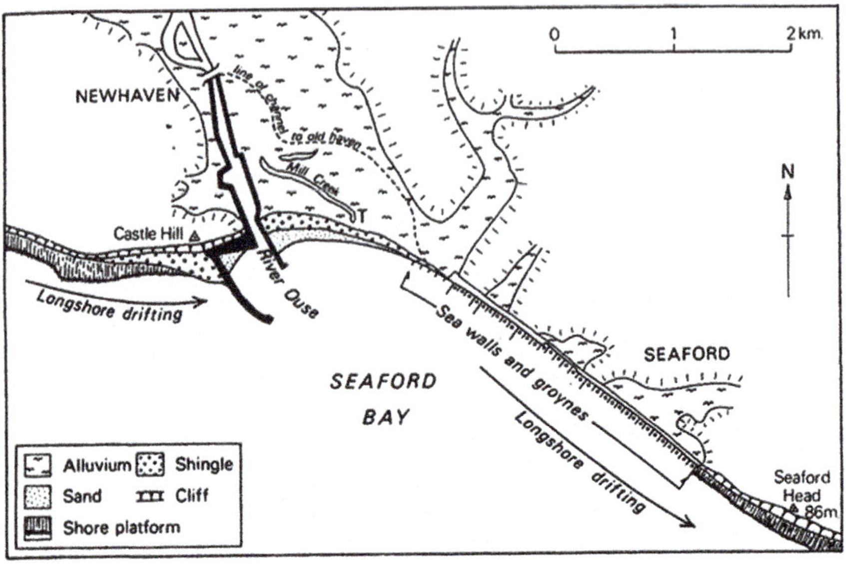

Fig. 4.14 The shingle beach at Seaford in Sussex was supplied with longshore drift until a harbour breakwater was built at Newhaven in 1845. Beach erosion ensued, and to counter this a sea wall and groynes were built along the Seaford esplanade. © Geostudies

Fig. 4.15 The renourished shingle beach at Seaford, Sussex in 2013. The renourished beach is retained by a breakwater at the eastern (downdrift) end. © Geostudies

blocks (Fig. 4.16), and to cover this with quartzite gravel brought by lorries from an inland quarry at Woodford, 12 km away: the natural beach had included sediment originating from coastal outcrops of this formation. The renourished beach was shaped as a terrace with a steep seaward slope. Two angled offshore breakwaters were then built to protect it from prevailing SW waves and storms, and a terminal groyne (East Pier) to prevent it being lost by longshore drift to the east (Fig. 4.17). The gravelly fill consisted of well-rounded but poorly sorted pebbles and cobbles, and initially had a pink colour from the earthy Triassic matrix. The beach was hosed down by the local fire brigade in an attempt to get rid of the pink stain, but it was more effectively removed by waves reworking the seaward slope, and did not last long. The beach restoration was completed in 1995.

The use of coarse sediment as a basis for a renourished sandy beach was also illustrated between Monte Circeo and Terracina, on the west coast of Italy.

Fig. 4.16 The sea wall at Sidmouth in South Devon was reinforced by dumping stone blocks after the bordering beach was depleted by erosion. Shingle was then added to restore the beach, which now looks similar to that at Seaford (Fig. 4.15). © Geostudies

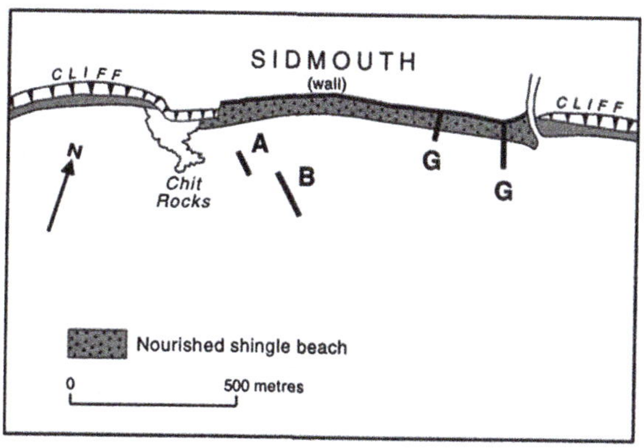

Fig. 4.17 When the shingle beach at Sidmouth in South Devon had been renourished, offshore breakwaters (*A*, *B*) were built to protect it from the prevailing SW waves and groynes (*G*) were inserted to retain it. © Geostudies

The beach was renourished in 1980–1983, using crushed limestone gravel to form a sloping terrace, which soon acquired a veneer of inwashed fine to medium sand. The coarse gravel thus provided a matrix for sustained sand accretion, as well as increasing shore protection (Evangelista et al. 1992).

Many seaside resorts have sought to renourish their beaches to improve the seaside environment. Cairns in north-eastern Australia has very little natural beach, the esplanade being fronted by a large area of mudflats exposed as the tide ebbs. In 1993 the beach was augmented by bringing 4,000 m^3 of medium to coarse sand excavated from a nearby delta, and depositing it between two temporary groynes in order to see whether a beach could be maintained along the muddy seafront. It is difficult to keep this beach clean, because it stands behind wide mudflats at low tide, and receives muddy sediment when it is washed by turbid water at high tide. Similar problems occur on the artificial resort beach of sand on Hamilton Island, Queensland, placed on a shore cleared of mangroves, and passing to mud at low tide.

The placement of an artificial beach where none existed previously occurred on the Algarve coast of Portugal (Fig. 4.18). The shore at Praia da Rocha was cliffed and mainly rocky, with only narrow pocket beaches, but in 1969–1970 880,000 m^3 of sand obtained from dredging the River Arade and excavation of the harbour at Portimão was pumped on to the beach, and a further 150,000 m^3 added from these

Fig. 4.18 At Praia da Rocha on the Algarve coast of Portugal an artificial beach has been placed in front of cliffs where no beach existed previously. Sediment dredged from the adjacent River Arade was deposited on the shore to the west in 1970 and 1983, forming a very wide beach above normal high tide level, incorporating several stacks. © Geostudies

sources in 1983. This formed an artificial beach 1,200 m long and up to 200 m wide in front of the cliffs and provided a beach for the seaside resort.

A similar project at Praia dos Três Castelos, to the west, failed when much of the 50–70 m wide beach emplaced in 1983 disappeared within 5 years, evidently because this sector was more exposed to wave scour than Praia da Rocha (Psuty and Moreira 1990).

Artificial beaches have been added on the seaward side of sea walls, as in the Netherlands, and harbour breakwaters, as at Cullen Bay, near Darwin in northern Australia. Here a beach of sand dredged from Darwin Harbour was placed in front of a large boulder breakwater built to enclose a marina in a former bay in 1993. This is a macrotidal coast, and the beach, emplaced around high tide level has been combed down by wave action on ebbing tides to a wide concave profile.

At Ediz Hook, on the southern shores of the Strait of Juan de Fuca, Washington, United States, a spit 5.6 km long and 27–275 m wide, shelters Port Angeles Harbour. It had been supplied naturally with sand and gravel from the west, partly from the Elhwa River and partly from the erosion of cliffs cut in glacial drift, but the damming of that river and the building of sea walls to halt erosion along the cliffed coast reduced the sediment supply, and beach erosion became severe. In 1977–1978 rock revetments were built, and it was decided to place gravelly material, quarried from glacial drift deposits west of Elhwa River and brought by truck, on the outer shore of the spit. Supplemented in 1985 by further such renourishment, the spit attained a relatively stable configuration (Galster and Schwartz 1990).

After the removal of dams on the Elwha River began in 2011 the supply of fluvial sediment to the shore resumed, with accretion on beaches near the river mouth.

Renourishment of beaches in south-east Queensland has been facilitated since 1988 by the use of the Port of Brisbane Authority dredge, modified to be able to pump sand out over its bow (the rainbow method, (Sect. 4.3.1, p. 57)) on to a beach. It was first used to place 50,000 m^3 of sand on an eroded beach at Woorim, on Bribie Island, north of Brisbane, and in 1992 to replenish Golden Beach at Caloundra, to the north, with 70,000 m^3 of sand dredged from nearby Pumicestone Passage. The rainbow technique has also been used to restore beaches in compartments between the numerous groynes on the shore at Felixstowe, on the east coast of England.

4.3.2 Emplacement by Longshore Drift

Losses of beach sediment alongshore are a common cause of beach erosion. Where longshore drift is unidirectional, or dominant in one direction, the losses can be made good by injecting sediment updrift and allowing it to spread along the eroded beach, as at Atlantic City, New Jersey, where sand deposited at the north-eastern end has been distributed south-west along the city seafront by wave action (Sect. 4.6, p. 87).

The idea of depositing a large quantity of sediment at a selected point on the beach and allowing wave action to spread it along the shore to renourish beaches downdrift was tested at San Onofre, California, in the 1980s. A 200,000 m^3 sand lobe was deposited, and surveys showed that at first there was some beach erosion downdrift, because the lobe was acting much like a breakwater, but this came to an end as the sand was distributed alongshore. The apex of the lobe migrated at about 2 m/day, and as it moved the sand lobe diminished rapidly in size, shrinking at the rate of about 50 % every 300 days, and becoming asymmetrical, attenuated and narrow as it prograded the beach downdrift (Grove et al. 1987).

More recently this concept has been applied on a large scale to the Dutch coast, where 21 million m^3 of sand were deposited in a hook-shaped lobe (Royal HaskoningDHV 2009; Stive et al. 2013). The aim was to produce growth of dunes and beach in the coastal section between Hoek van Holland and Scheveningen, increasing coastal protection from flooding as well as potential for habitat creation and recreation development. This method of design has been termed a 'sand engine', but this term should be used with caution as it could imply that there is a mechanical component to the concept. There is in fact no mechanical placement or modification of the of beach sediment following the initial placement. The main expectation of a 'sand engine' is that the placed beach sediment will stabilise the coastline in its present position and feed sand about 10 km downdrift over an extended period (20 years) as the emplaced sediment is shaped by waves, wind and currents.

Results of numerical modelling and observed data from the first year were presented by Stive et al. (2013). This included the stormy winter of 2011–2012, when the shape of the deposited sand peninsula changed considerably (Fig. 4.19). The maximum width decreased from 0.96 to 0.84 km, while its length alongshore increased from 2.4 to 3.6 km. The sediment volume at the location of the initial peninsula decreased during this first year by about 1.4 million m^3, while adjacent coastal sectors showed an increase in sediment volume of 0.9 million m^3, confirming the longshore distribution.

Renourishment of the beach on the German North Sea island of Sylt began in 1972, using 770,000 m^3 of sand obtained by shallow dredging and deposited as a large lobe protruding from the shore. This sand was gradually distributed downdrift by wave action. Progress was monitored on profiles surveyed at 500 m intervals, which showed that after 5 years more than 60 % of the sand deposited in the lobe had moved on to the beach downdrift. On the basis of this experience it was decided that the optimum site for lobe deposition should be a kilometre updrift of the site chosen initially. The project showed that renourishment by means of redistribution from a deposited lobe was feasible, and that the location of such a lobe should be well updrift of the sector to receive the beach renourishment. A similar principle guided the dumping of urban rubble on the shore near Odessa to renourish beaches downdrift.

Several beach renourishment projects have used the principle that longshore drift interrupted by a tidal inlet with strong transverse ebb and flow currents (which act like a breakwater) can be restored by sealing off the inlet or cutting a new one. This has been illustrated on Seabrook Island, South Carolina, where beach erosion became severe when southward drifting of sand was impeded by Captain Sams Inlet, with interception on the northern (updrift) side. In due course the sand accumulating

Fig. 4.19 A time series of oblique aerial photographs taken across the 'sand engine'. Photographs show morphological development since completion in July 2011. From *top* to *bottom*: 5 July 2011, 13 October 2011, 30 March 2012 and 4 September 2012. © Rijkswaterstaat (The executive branch of the Dutch Ministry of Infrastructure and the Environment)/Joop van Houdt

on the updrift side began to form a spit, which grew southward, deflecting the mouth of the inlet, and it was decided to cut a channel through this, and allow 170,000 km^3 of sand to drift on southward. The released sand soon sealed the former inlet, and moved on to renourish the previously eroding beach, widening it by more than 300 m over the next 6 years, south of the new artificial cut (Kana 1989).

Port Phillip Bay, Australia, is a marine embayment with a narrow (3.2 km) entrance from Bass Strait to the south, a spring tide range of only about 60 cm, and weak nearshore tidal currents. Seasonally alternating longshore drift here is seen on the east coast of this bay as a dominance of northward drift in the summer half-year (November to April), lowering and narrowing beaches at their southern ends and widening and raising them at their northern ends. In the winter half-year, between May and October, this is reversed when the dominant waves arrive from the NW, drifting beach sediment southward. The sequence is illustrated on Black Rock beach (Fig. 4.20).

Beach renourishment projects here must be designed to allow for the seasonal alternation and the possibility that the gains at each end of a beach compartment

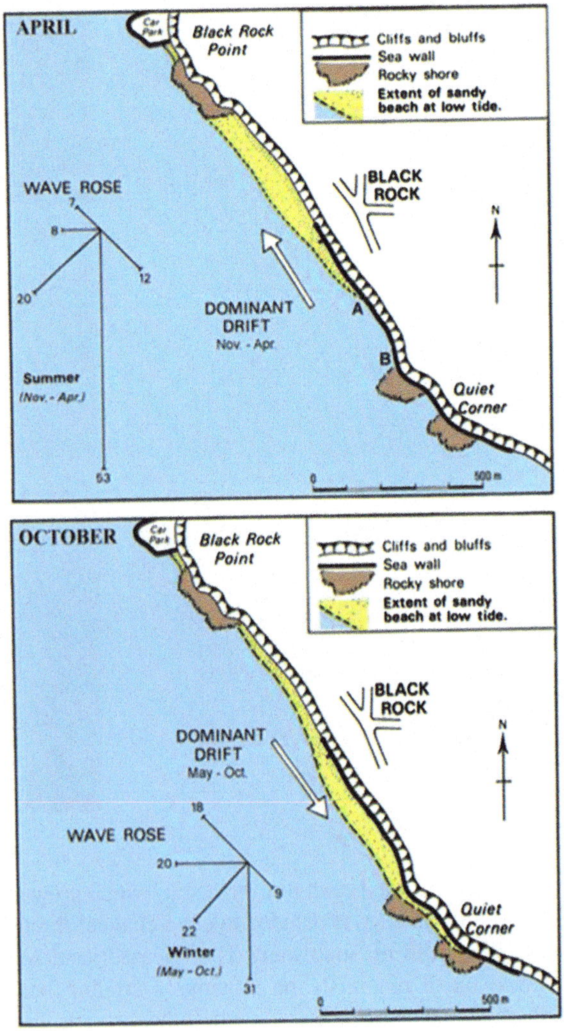

Fig. 4.20 Dominant longshore drift on the beach at Black Rock, on the NE coast of Port Phillip Bay, Australia, is northward during the summer half-year (November–April) and southward during the winter half-year (May–October). © Geostudies

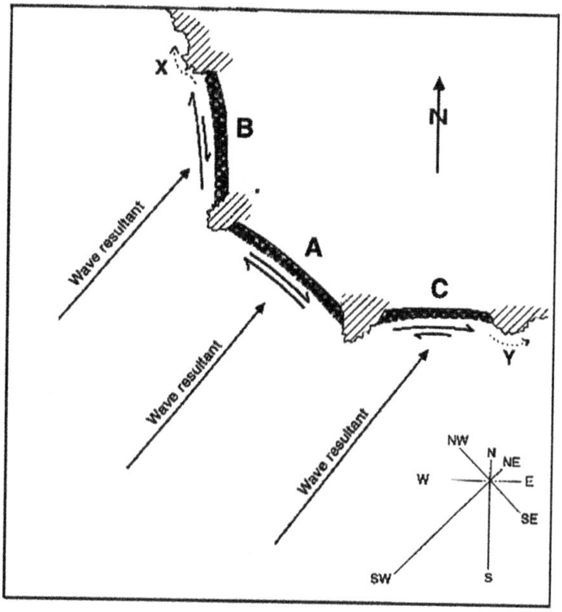

Fig. 4.21 Variation in net longshore drift with aspect: *A* balanced longshore drift when the wave resultant is orthogonal to the shore, *B* net northward drift on a west-facing beach and *C* net eastward drift on a south-facing beach when the wave resultant arrives from the SW. © Geostudies

will result in losses past headlands or breakwaters. In these circumstances, coastline orientation in relation to wave climate determines whether an emplaced beach will remain in position, or lose sediment in one direction or the other (Fig. 4.21), so that there is a contrast between predominantly northward drift on Hampton Beach (facing south-west) and predominantly south-eastward drift on Quiet Corner Beach (which has a more southerly aspect) a few kilometres away (Bird 1991).

4.3.3 Bypassing

Bypassing is the passage of sediment past an obstacle such as a tidal inlet or breakwater to replenish a beach downdrift. It occurs naturally, but can also be developed artificially as a means of downdrift beach replenishment. It has been used at several places worldwide (Boswood and Murray 2001), but has proved particularly popular on the coasts of the United States and Australia. An early example of beach renourishment using bypassing as a means of artificially restoring the longshore drift regime was at South Lake Worth Inlet in Pam Beach County, Florida. After this tidal inlet was stabilised by breakwater construction in the 1930s to improve navigation the nearby beaches soon showed updrift accretion and downdrift erosion. When groynes failed to control the downdrift erosion a sand bypassing scheme was introduced in 1935, taking about 48,000 m^3 of sand per year from the accreting southern beach round to renourish the depleted northern beach. This was the longest continuously operating fixed sand by-passing plant in the world (Finkl and Walker 2004). Similar projects have been used at Santa

Barbara, Ventura, and Channel Islands Harbour in California, both of which had accretion updrift and erosion downdrift of harbour breakwaters.

There are many techniques of sand bypassing (Bruun and Willekes 1992; USACE 1991; Bruun 1996; Boswood and Murray 2001). Hydraulic methods of both sediment collection and delivery have proven to be one of the most effective contemporary methods for sand bypassing (Boswood et al. 2005; Mocke et al. 2005; Acworth and Lawson 2011).

At the entrance to the Tweed River on Australia's Gold Coast the Tweed River Entrance Sand Bypassing Project (TREBP) was established in 2001 to mitigate the substantial decrease in northward drift intercepted by the Tweed River Entrance training walls. A permanent sand intake jetty intercepts sand moving northward, which is then piped under the Tweed River at a rate consistent with the average natural longshore transport (of the order of 500,000 m^3), then released on the beaches north of the Tweed River Entrance. Previous to the TREBP intermittent sand nourishment was undertaken using offshore sand reserves (Boak et al. 2001). The TREBP has been successful in changing the morphology of the beaches north of the entrance (Castelle et al. 2006) providing a wider beach, better amenity, a demonstrated storm buffer and surfing benefits (Boswood et al. 2005).

As a result of a combination of weir jetty, interior sand trap and dredging, the Hillsboro Inlet in Broward County, southeast Florida, has been able to bypass 100 % of the estimated net longshore drift since 1965, when this was the first sand by-passing weir jetty in the world (Finkl 1993).

An alternative is to carry sediment round past a harbour entrance in trucks, as at Hvidesande in Denmark, or ferry it past the entrance, as at Port Hueneme, California. However, some eroding beaches downdrift of breakwaters have been renourished with sediment from other sources, notably the sea floor, as at Seaford in England, Timaru in New Zealand and Lagos in Nigeria.

The Lagos coastline has been much modified by the building of breakwaters to stabilise the harbour entrance. These caused updrift accretion on Lighthouse Beach and downdrift erosion on Victoria Beach (Usoro 2010). To counter erosion Victoria Beach was renourished with sand pumped in from the sea floor in 1976, and it is now accepted that this beach will need to be replenished frequently, perhaps with the aid of a sand bypassing system from the accreting Lighthouse Beach, west of the breakwaters (Ibe et al. 1991).

4.3.4 Recycling

Beach sediment carried along the coast by longshore drift can be brought back to renourish an eroded beach, a process termed recycling (Willis and Price 1975).

The shingle beach at Rye on the Sussex coast in England is subject to longshore drift, which has depleted the Cliff End beach at the updrift end. The supply of shingle to this beach has been reduced because recurrent landslides on the Fairlight coast to the west interrupted longshore drift from Hastings. Stabilisation of the mouth of the River Rother downdrift to the east by breakwaters built in the 19th century began

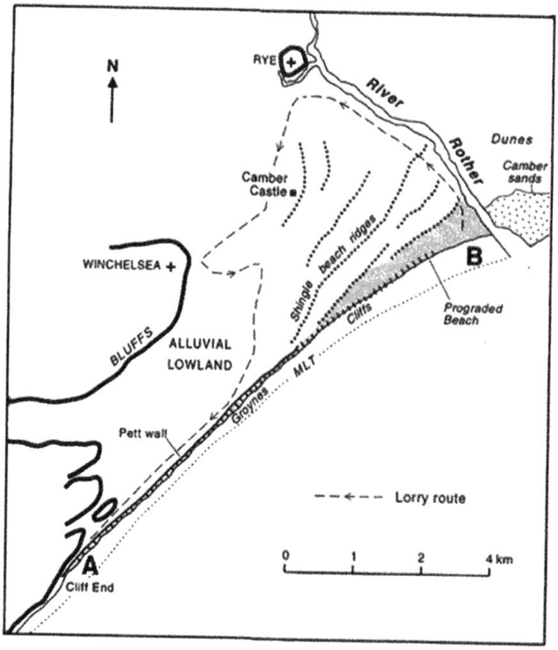

Fig. 4.22 On the coast near Rye in Sussex longshore drift carries shingle from Cliff End (*A*) to accrete alongside the River Rother breakwater (*B*), and from here it is recycled by lorries that take it back to Cliff End. © Geostudies

to intercept this drifting shingle and form a wide beach. In 1934 some shingle was taken from the accreting area adjacent to the Rye breakwater and returned to the depleted beach to the west, using a beach railway. In 1955 the Kent River Board began regular ferrying of lorry loads of shingle taken from alongside the breakwater round to the western end and dump them to restore the depleted beach (Fig. 4.22): in the first year 155,000 m^3 of shingle was returned to the SW end of the beach, and in subsequent years similar quantities were recycled (Eddison 1983).

On the Adelaide coast in South Australia sand lost from the southern beaches has drifted north towards Port Adelaide, where the wide beach has been used as a source of sand trucked back southward to the eroding beaches (Fig. 4.23) (Wynne 1984). Since 2010 it has been found more practical to recycle sand on a smaller scale within selected beach compartment (Fig. 4.24a, b).

At Noosa, on the east coast of Australia, a hydraulic system of recycling was installed in 2004 to replace previous recycling efforts which included dredging of Noosa Sound. The hydraulic system collects sand from the downdrift end of the beach and pumps it back along the beach to provide beach nourishment (Nankervis 2005). The system is permanently installed and operated regularly to provide a constant supply (35,000 m^3/year) of sand to Noosa Beach.

Alternatively, recycling can be carried out by barges travelling along the shore, as on the Caucasian Black Sea coast (Sect. 4.2.4, p. 47) or by pumping the sand back from the updrift end of the beach through a pipe, as at Narva Bay in Estonia. Recycling has the advantage that the shape and size characteristics of the sediment taken from downdrift will be similar to those on the depleted updrift beach, but

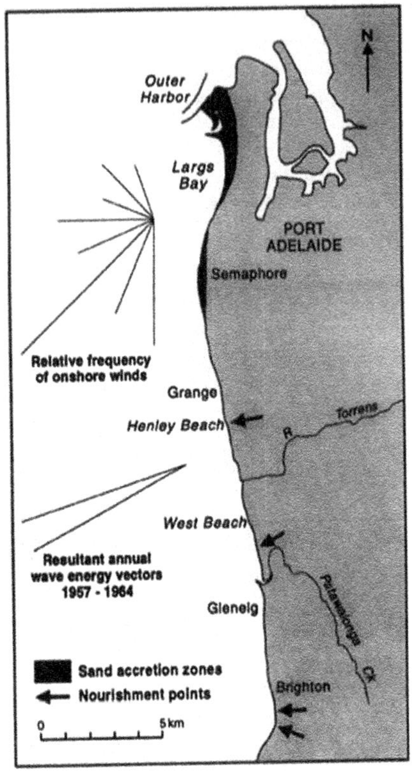

Fig. 4.23 On the Adelaide coast in South Australia beach sand drifts northward to accumulate at Semaphore and in Largs Bay. It is recycled by lorries to nourishment points (*arrowed*) to the south. © Geostudies

there will be gradual attrition, and supplementary coarser sediment may need to be imported eventually.

In addition to recycling of sediment carried along the coast by longshore drift, sediment removed from a beach and deposited offshore can also be recycled. At Ettalong Beach, on the north side of Broken Bay, New South Wales, ocean swell and tidal currents transport sand alongshore as far as a change in the orientation of the coastline, where it is deposited in offshore shoals. Sand from these shoals is periodically dredged and returned to Ettalong Beach. The latest renourishment was undertaken in 2013 when 30,000 m^3 of sediment was dredged from the sea floor and pumped back to the beach.

4.3.5 Use of Shoreward Drift

On swash-dominated beaches there is the possibility of dumping sediment in the shallow nearshore area and allowing swash to deliver it to the beach. Studies of onshore-offshore movements of sediment in relation to wave and current regimes (usually seaward movement during storms and shoreward movement during

Fig. 4.24a Northward drifting sand on the Adelaide coast trapped by a breakwater near West Beach is loaded into trucks and carried southward along the beach to Glenelg (Fig. 4.24b). © Geostudies

calmer weather) are necessary to determine the depth from which such shoreward transportation will occur, and therefore the optimum location for dumping sediment to be washed on to the beach.

The proportion of sand dumped in the nearshore zone that moves up on to the beach varies with wave conditions. In North Carolina Schwartz and Musialawski (1977) found that up to 75 % of dredged river sediment dumped in the nearshore zone was washed on to the beach, but if stormy conditions followed the nearshore dumping little or none of it moved onshore. There is usually also some longshore drift, which will determine which sectors of the beach actually receive the inwashed sediment.

At New River Inlet on the coast of North Carolina, an attempt was made to renourish an eroding beach that had formerly received sediment from the sea floor by dumping sand nearshore in the expectation that it would be washed on to the beach by wave action. 26,750 m^3 of coarse sand dredged from New River was dumped on the sea floor by split-hull barges, and its movement followed by monitoring beach and nearshore profiles. The study showed that the sand deposited in depths of less than 4 m moved shoreward over the ensuing 13 weeks, whereas sand deposited at greater depths moved seaward. The sand that moved shoreward was deflected alongshore by obliquely-arriving waves to beaches down the coast

Fig. 4.24b Sand heaps delivered to the depleted southern end of the beach at Glenelg, Adelaide for beach renourishment. © Geostudies

(Schwartz and Musialawski 1977). The project indicated that if sufficient sand is deposited in the nearshore zone it will move shoreward from a specific depth, but that longshore drift will determine which sectors of the beach receive it.

On the Gold Coast in Queensland, Australia 100,000 m^3 of sand was dredged from the sea floor 1.5 km off Burleigh Heads (where the water is 18–25 m deep) in 1985, brought in and dumped on sand bars in the nearshore zone. Within a year much of this sand had been washed on to the beach.

A similar method was utilised in the Hythe Coast Protection Scheme in Kent,United Kingdom where it was demonstrated that by placing the replenishment sediment in one location on the beach face and allowing the beach to sort naturally, without labour-intensive re-profiling, the duration (construction) and cost of recycling was significantly reduced (Clarke and Brookes 2008).

4.3.6 Backpassing and Beach Re-shaping

Losses of sediment seaward from a beach, particularly during stormy phases, can be offset by backpassing, the retrieval of beach material that has been swept offshore and its return to the beach. This is analogous to recycling of beach material carried away by longshore drift, mentioned above, and is important in beach profile renourishment, discussed below.

Backpassing is a possibility where wave energy is low, especially if there is a wide intertidal area, as on the southern shores of Port Phillip Bay in south-east Australia, where a long, gently curving sandy beach, much used by summer holidaymakers, extends from Rosebud to Rye. It is fronted by multiple parallel sand bars that run out seaward across a shallow intertidal area up to 500 m wide, and move to and fro in response to alternations of obliquely-arriving wave action. In the 1950s erosion of the beach prompted the building of sea walls and groynes, but depletion continued, and in 1963 it was decided to renourish one sector by bulldozing sand in from the nearshore sand bars at low tide. This was successful, and during the next 20 years several sectors of the beach were built up and widened in this way. Between 3,000 and 5,500 m^3 of nearshore sand were delivered to the beach annually, and parts of the nearshore area deepened by up to 30 cm as the bulldozer scooped sand shoreward. However, there was an ensuing problem of dense seagrass infestation in the areas deepened by bulldozing, and the beach and nearshore area had to be restored by renourishment. This was achieved in 1985 by dumping a series of artificial transverse bars of fine sand 5 m wide and 120 m long, spaced at 100 m intervals, which were widened and moved to and fro by wave action until they buried and destroyed the nearshore seagrass beds. Within 2 years the artificial transverse bars had been re-shaped by wave action, and their lateral migration had reduced the seagrass area to a few small patches amid the distributed sand (Parry and Collett 1985).

Several beaches on the coast of England and Wales have been modified by the combing of shingle down the beach by plunging waves during storms. This can be countered by bulldozing gravel up to the back of the shore to raise the upper beach profile in front of a sea wall or eroding cliff. This has been effective at Dunwich, East Anglia (Fig. 4.25), as a short-term procedure, where the aim is to increase upper beach protection of soft cliffs cut in glacial drift. Where storm downcombing becomes frequent it may be necessary to add more sediment in order to renourish a higher and wider beach that is more protective.

Fig. 4.25 Apron of shingle bulldozed to form an upper beach at Dunwich on the Suffolk coast. © Geostudies

After storm surge flooding overtopped Chesil Beach at Chiswell near Portland in the 1970s the beach crest was raised by bulldozing up shingle, and then stabilised with a capping of gabion mattresses (caged stones). Where beaches lose sediment by overwashing during storms it can be bulldozed back from the inner slope to the beach face. This has been successful on the shingle barrier beach north of Timaru, on the Canterbury coast, South Island, New Zealand, where sediment washed over into Washdyke Lagoon by storm waves was retrieved in this way (Kirk and Weaver 1985).

A different kind of backpassing may be necessary where wind action carries sand to the back of the beach. At Harrison County, Mississippi sand blown from the renourished beach by occasional strong wind action during hurricanes piles up against the wall to the rear, and has to be periodically taken back by trucking to restore the beach profile.

Many seaside resorts improve their beaches for summer holiday use by sweeping back sand and shingle dumped on their esplanades by winter storms. Some, such as Weymouth in southern England, take care to maintain a clean, flat sandy beach as an attraction for children and a venue for sand castle competitions each summer. Each winter, sand is washed and blown round the shore of Weymouth Bay to accumulate as low dunes in front of the sea wall at the southern end of the esplanade, and in spring this is collected and redistributed across the beach.

4.3.7 Overfill

It is generally necessary to add more beach material than is necessary to restore a beach to its natural dimensions, in order to allow for expected subsequent losses onshore, offshore or alongshore. It is common practice to use overfill ratios, as recommended by the Coastal Engineering Manual (USACE 2002) to compensate for the rapid removal of finer material, by simply adding more.

James (1975) defined the Overfill Ratio as the volume of sediment necessary to restore a beach similar to the natural beach, allowing for losses of sediment until the grain size had been sorted to match the natural distribution. The Renourishment Factor is the ratio of the rate of erosion of beach fill material to the preceding rate of natural beach erosion, indicating the frequency of replenishment necessary to maintain a stable beach volume. Overfill is also necessary to anticipate losses due to spilling of sediment out of the renourished area, around a terminal groyne or bordering headland or terminal groyne. Renourished Dutch beaches are usually overfilled, with overfill volumes typically varying between 10 and 40 % (Hanson et al. 2002; Verhargen 1992).

4.3.8 Shore Profile Renourishment

Most beach renourishment projects form a beach terrace, which is then re-shaped by waves and currents towards a natural concave profile, often with sand bars just offshore. It has been suggested that it may be more useful to renourish the whole profile, including backshore dunes and the nearshore sea floor, and not just the upper beach terrace, in order to attain a more stable transverse configuration (Bruun 1990). Shore profile renourishment is preferred to upper beach renourishment because the latter leaves unnaturally steep seaward edges, which can be reflective and a cause of nearshore scour, and are also subject to erosion and reshaping, with often rapid initial losses. Profile renourishment reduces these losses, and has lower construction and maintenance costs. It also permits the use of a wider range of grain sizes, subject to sorting by wave processes into appropriate zones on the profile.

The aim is to establish a relatively stable 'equilibrium profile'. Beaches with concave profiles are more stable than those with straight, convex or irregular profiles, and once concave profiles are attained they become relatively (but not absolutely) stable. The gradient of the concave profiles varies with grain size and preceding wave conditions. Subsequent oscillations occur with episodes of storm wave erosion and fine weather accretion. As long as the beach profile oscillates between these limits the beach can be considered relatively stable, but as has been noted (Sect. 1, p. 2) beach erosion is prevalent and most renourished beaches need further inputs of beach sediment.

In the Netherlands the value of backshore dunes as a reservoir of sand and a barrier to storm waves and marine flooding has long been realised. On the Atlantic coast of the United States, Kana and Stevens (1990) discussed techniques of beach and dune profile restoration following erosion by a hurricane.

Bruun (1990) noted that shore profile renourishment required dredging and dumping equipment of the kind used in the Netherlands (Stive et al. 1991). A transverse profile can be maintained by backpassing (Sect. 4.3.6, p. 70), using permanent offshore dredging and pumping stations, but this could lead to frequent disruption of sea floor plant and animal communities and fish habitats, and some would consider such offshore structures obtrusive.

Renourishment of the whole shore profile was carried out after the failure of several projects that dealt only with the upper beach terrace at Ocean City, New Jersey. Detailed investigations prepared the way for a project, which used 4.6 million m^3 of sediment to shape a beach 30 m wide, with a concave shore profile on the seaward side, as well as a backshore dune (Fulford and Grosskopf 1989; Anders and Hansen 1990). Subsequent changes were monitored, and in January 1992 a major storm removed most of the beach and part of the dune (Houston 1995), but there was little property damage in the resort, and it seemed likely that a protective beach and dune could be maintained if the beach profile were renourished after each storm. Renourishment of the whole of the shore profile, including the nearshore zone and backshore dunes, was thus seen as a more effective way of establishing and maintaining a protective beach than simply dumping sand to form an upper beach terrace.

At Miami the beach renourished in 1975 included a low backshore dune, a flat terrace and a gentle seaward slope, a landform association that has remained fairly stable, proving remarkably resilient even during successive hurricanes (Finkl 1981).

In the decade following 1975 10 million m^3 of sand was pumped ashore along 17 km of the coast to form a beach over 150 m wide. The aim was to restore the beach and provide a formation that would protect the coastline.

In the Netherlands, a region of subsiding coastal land, the response to the gradual rise of sea level over many centuries has been to counter-attack by building and enlarging sea walls (dykes) to keep out the sea. Extensive areas of land are now below high tide level. Some of the sea walls enclose tidal marshlands and shallow sea areas for land reclamation. Large sea walls now dominate long sectors of the Dutch coastline, and in places beaches have been added on their seaward side, as at Neuharlingersiel, East Friesland (Jelgersma 1975). Where beaches and dunes still exist the aim of Dutch engineers has been to maintain the coast by renourishing the nearshore zone as well as the beach, and providing sufficient sand to maintain a beach profile that prevents erosion of the dunes behind the beach (Roelvink 1989). It has been calculated that between 6 and 10 million m^3 of sediment fill will be required annually to maintain the existing coastline as subsidence continues (Louisse and Kuik 1990). Where necessary the coast will be built forward to reduce impacts of future erosion, bearing in mind the probability of an accelerating sea level rise in the next few decades (Pluijm 1990).

If a sufficiently wide beach is formed, wind and wave action will then shape a shore profile that includes backshore dunes as well as nearshore sand bars. At Noosa in Queensland, Australia, beach erosion followed diversion of the Noosa River outlet (Coughlan 1989). Sand sprayed by the rainbow technique on to the shore in Granite Bay drifted round to renourish the main beach at Noosa, where it was retained by a groyne. In due course it prograded to form a wide area of bare sandy beach, the landward part of which was shaped by wind action into backshore dunes, now stabilised by the planting of trees and shrubs.

4.3.9 Part-Renourishments

Beach renourishments take the form of partial fills, either as a layer of sediment placed on top of an existing beach or as sediment placed under an existing beach. For example in the latter case, filled geotextile bags have been used in the United Kingdom, placed under existing beach material to raise the berm and provide an increased level of stability. Although not technically classed as beach renourishment, beach sediment is added to the beach. The use of geotextile bags as opposed to loose sediment provides a more durable feature.

More common is the addition of renourishment sediment placed on top of the existing beach material. At seaside resorts the losses of sandy sediment during winter storms may leave a gravelly beach, and it is then necessary to bring back sand to restore the beach for recreational use. At Whitburn Bay on the Durham coast in north-east England the resort beach is improved each spring by dumping sand over the gravelly shore (Fig. 4.26). This is termed a veneer, as when sand is deposited over much coarser (boulders coral rocks) or finer (silt and silty sand)

Fig. 4.26 Improvement of the recreational beach at Whitburn Bay, Durham, by dumping sand over shingle. © Geostudies

sediment (Finkl and Walker 2004). Veneer renourishment has been used at Corpus Christi, Texas and Grand Island, Los Angeles, USA (NRC 1995).

4.3.10 Use of Structures

It is common for beach renourishment projects to include supporting structures in an attempt to reduce the rate of losses and improve performance. For example, it may be necessary to build groynes to retain a renourished beach on a coast dominated by longshore drift. Sand and gravel drifting eastward from Bournemouth augmented beaches as far as Hengistbury Head, where a terminal groyne was built to intercept the drifting sediment to widen the beach and protect the backing cliffs from further erosion (Fig. 3.4).

In Monterey Bay, California, sand from a quarry was dumped on the shore alongside a retaining groyne built at Capitola to prevent losses downdrift (Griggs 1990). Reference has been made to groynes built to retain renourished beaches at Seaford in England and Mentone, Brighton and Sandringham in Port Phillip Bay, Australia.

When shingle dredged from the sea floor was brought into renourish a depleted beach in front of the sea wall at Lodmoor, east of the seaside resort at Weymouth in Dorset in 1995, there was a possibility that during south-easterly storms shingle from the augmented beach would drift westward on to the sandy resort beach. In order to prevent this, a T-shaped retaining groyne was constructed at Melcombe Regis, at the western end of the renourished beach.

The disadvantage of groynes is that they generally result in erosion downdrift. On Sandy Hook, New Jersey, the shore had been protected by an 11 km sea wall, to which groynes were added, but erosion beyond the end of these structures began to cut out a bay. In 1977 152,920 m^3 of sand was trucked in and deposited in the bay to provide protection from storm damage, but it soon drifted away northward (Nordstrom et al. 1979). It would have been possible to go on extending the groyne field downdrift, but Nordstrom and Allen (1980) suggested it might be preferable to abandon the groyne field and provide a supply of sand at the updrift end sufficient to maintain a protective beach along the coast by longshore drift.

Groynes are of little use where sediment from the renourished beach between them is being withdrawn to the sea floor, as at Virginia Beach on the Atlantic coast of the United States, or the German North Sea island of Nordeney (Kunz 1990). In such conditions it may be possible to retain a renourished beach by building nearshore underwater breakwaters to prevent seaward losses, as at Niigata in Japan (Chill et al. 1989).

Artificial structures such as marinas may act as traps for seasonally drifting sediment, as at Sandringham Harbour on the NE coast of Port Phillip Bay, Australia. When the first survey was carried out in 1861 this beach compartment had a low receding cliff in soft Red Bluff Sand bordered by a sandy beach that drifted to and fro seasonally. Cliff recession prompted the building of a sea wall in the late 1930s, when the cliffs were graded back to an artificial slope and planted with vegetation. As a sequel, the beach between Green Point and Hampton was depleted, because it was no longer maintained by the supply of sand washed out of the receding cliff by runoff and wave scour. This beach was further reduced following the building of a boat harbour (Fig. 4.27a), with construction and elaboration of a large boulder breakwater at Picnic Point, to the south (Fig. 4.27b). As on the other beaches on the north-east coast of Port Phillip Bay, there is northward drift of beach sand between November and April, and southward drift from May

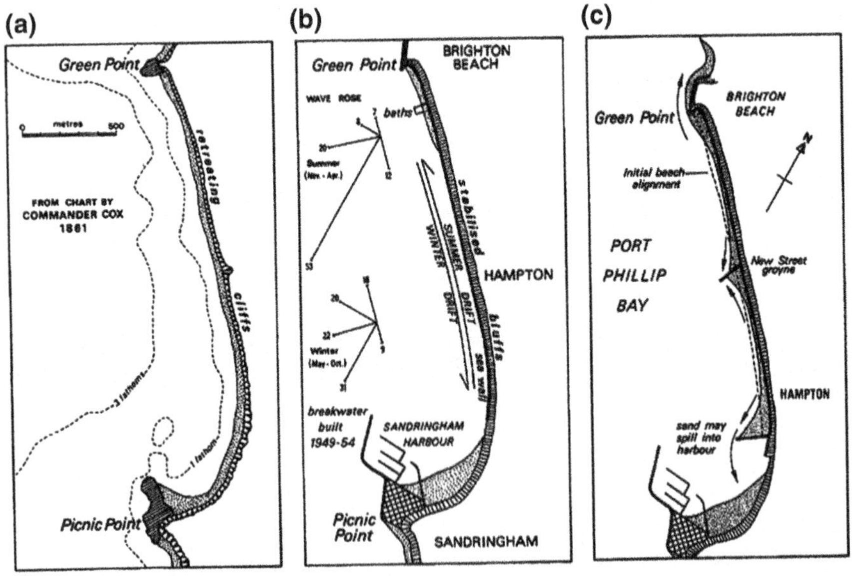

Fig. 4.27 Changes on the coast at Sandringham, Port Phillip Bay Australia. **a** The natural configuration in 1861, **b** seasonal alternations of longshore drift on the beach resulted in its depletion after a breakwater was built, and southward-drifting sand accumulated in Sandringham harbour, **c** renourished beaches are now maintained between groynes. © Geostudies

to October. After the harbour breakwater was completed in 1954 it acted as a trap for beach sand drifting southward each winter, and prevented it from being carried back by south-westerly waves in the summer months. By 1960, little beach sand was left on the Hampton coast, much of it having drifted into the lee of the Sandringham Harbour breakwater to accumulate as a wide prograded sandy plain (Bird 1996). This reduced the area of the harbour, which was also shallowed by sand deposition.

Depletion of the beach between Green Point and Hampton left the sea wall exposed to damage by storm waves, and the Victorian Division of Ports and Harbors then decided to renourish the northern part of the beach. In order to prevent sand drifting south in the winter months into Sandringham Harbour it was necessary to construct a boulder groyne 160 m long in the middle of Hampton Bay at an angle of 65° to the coastline, so that the emplaced beach was exposed to south-westerly wave action, which could drive the accumulated sand back northward in summer. Early in 1987 108,000 m^3 of coarse sand was dredged from an area 2 km offshore and piped into form a beach 40 m wide and 1,100 m long, extending from this groyne north to Green Point, known as South Brighton beach (Fig. 4.28); it was initially formed as a terrace 2 m above low spring tide level, the top of the groyne being 0.3 m higher.

In the winter of 1987 some of the sand was washed southward over the groyne (Fig. 4.28), and in the following summer the beach towards Green Point was widened by northward drift, with some sand moving past this headland. Over the next few years this sequence was repeated, and successive spits were built out north from Green Point in summer, pushed back by storm waves in the following winter

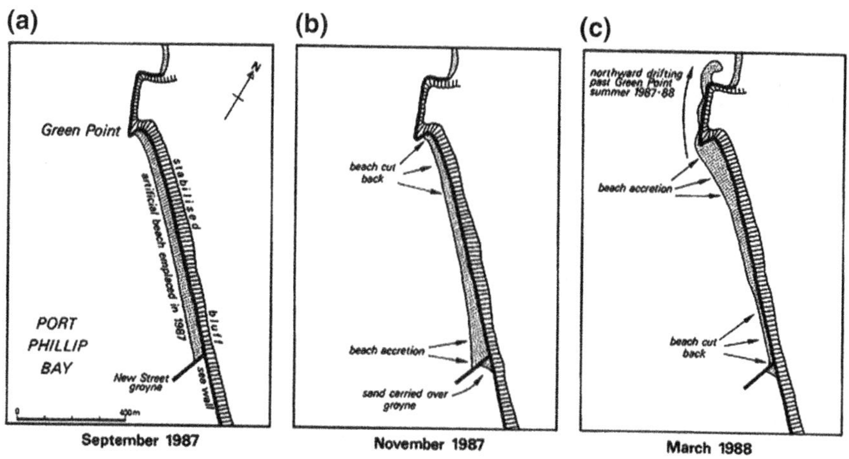

Fig. 4.28 At Brighton on the NE coast of Port Phillip Bay an artificial beach was inserted between Green Point and a groyne constructed at New Street in 1987 (**a**). Some of the sand was washed southward over the groyne (**b**) during the following winter, and some drifted round Green Point to form a spit to the north (**c**) in the following summer. © Geostudies

and added to the next beach. Eventually the losses of sand depleted the renourished beach until it no longer prograded sufficiently in summer to leak past Green Point, and by 1994 it had become fairly stable, apart from continuing seasonal alternations within the beach compartment. The beach has become narrow at the southern end.

Two more groynes were added south of New Street in 1996, and intervening beaches then renourished. These beaches have persisted, and show that a renourished beach can be maintained when a beach compartment is divided into segments by building groynes.

Offshore breakwaters have been used to create a pattern of refracted waves that will concentrate sand deposition and prograde the beach in the lee of the breakwater. This has been illustrated at Santa Monica, California.

At Port Hueneme in California, a breakwater was built parallel to the coast on the updrift side of the harbour entrance in 1940, and the sandy cusp that formed on the beach landward of it (Fig. 4.29) has been excavated at the rate of about 400,000 m^3/year by a floating dredge to produce sand which is used to replenish wasting beaches on the downdrift shore (Johnson 1959).

It has been suggested that a floating breakwater, anchored off successive sectors of the shore, could induce local accretion of sand and gravel by shoreward drift of sediment to renourish a beach in stages along the coast. Reference has been made to the use of submarine breakwaters to diminish wave scour and protect a renourished beach at Niigata, Japan. At Marina di Cecina, on the Tuscan coast, Italy, beach renourishment was accompanied by the building of retentive groynes with undersea extensions. These have been successful, but two of the submerged breakwaters caused current scour leading to a sediment deficit and some beach erosion, and these are being allowed to disintegrate (Ciprani et al. 1992).

Fig. 4.29 Sand by-passing at Port Hueneme, California. © Geostudies

4.4 Design Considerations

4.4.1 Introduction

There are a number of significant specific design considerations when planning beach renourishment. In many cases, design will be limited, but it can be complex, assessing a range of criteria in detail. For example, beach renourishment design can vary from basic forms, such as the placement of sediment on the subaerial beach re-using dredged sediment, through to complex design, considering multiple factors such as the local rate of erosion, the natural beach slope, native and imported sediment grain sizes, wave climate and tidal levels, existing structures and infrastructure, and past engineering activities in the area. In addition the timing and effects of multiple renourishments can be considered (Dean 1995). Design considerations concerning grain size and grading, profile shape and volume, and planform shape (all strongly interlinked) are now summarised. Detailed discussion of all design considerations is beyond the scope of this textbook, but further technical details are available in design manuals (Sect. 1, p. 3, 5).

Where possible, beach renourishment design should take as much information as possible from the existing site, including erosion history. Mention has been made of the Dutch Method of design (Sect. 4.3.7, p. 72), described by Verhargen (1992). This is used in most beach renourishment projects in the Netherlands, and follows five steps: (1) regular (at least annual) surveys of beach profiles over a period of at least 10 years; (2) calculation of annual sand loss in each coastal sector; (3) addition of 40 % to offset losses as the profile adjusts to greater beach width extending into deeper water; (4) multiplication of this quantity by an duration; and (5) placing of sediment between the dune foot and a line 1 m below low tide.

4.4.2 Grain Size and Grading

It is necessary to consider the calibre of the sediment needed for a beach renourishment (USACE 2002) because the median grain size and sorting of sediment influences the stability of the renourished beach (CIRIA 2010). It has long been known that beach renourishment is most effective when an appropriate grain-sized sediment is used (Newman 1976). Sediment used for renourishment should have similar grain size characteristics to the natural beach, for excessively fine sediment is soon lost offshore or alongshore and excessively coarse sediment may form too steep a beach, promoting nearshore reflection scour (Krumbein 1957) and developing a beach profile unsuitable for recreational use (Blott and Pye 2003). In general, coarser sediment is likely to persist longer on a beach, and finer sediment to be lost more quickly (Dean 1983; Bird 1996; Nordstrom 2000; Komar 2007; Kumada et al. 2007). This has led to a number of renourishment projects incorporating coarser sediment than the native sediment. Monitoring of renourished beaches on the shores of Lake Michigan, in the United States, has shown that those formed of gravel (pebbles or cobbles) last about five times as long as those of fine sand, which disappears quickly offshore (Roellig 1989). Practical beach experiments were conducted at Tankerton in southeast England to monitor beach evolution. Five compartments between groynes were filled with significantly different material: locally recycled, very fine, very coarse, an experimental cap and a standard replenishment mix (Clarke and Brookes 2008). The results demonstrated a continued loss of fine sediment, with over double the amount removed when compared to the coarser material. However, there is marked reluctance of people used to sandy beaches to have them replaced by more stable, coarser shingle.

In addition to median grain size, grading can affect the morphology (especially profile) of the beach, which is related to porosity and permeability (She et al. 2007).

Rapid losses have occurred on beaches renourished with unsuitable sediment. In 1982 the 12 km beach fronting the seaside resort at Ocean City, New Jersey, was renourished using poorly sorted shelly sand dredged from a nearby tidal delta, but this was quickly washed away by storm waves (Pilkey and Clayton 1987). An example of the successful use of coarse sediment after finer sediment had been washed away occurred on the Presque Isle Peninsula on the Pennsylvania shore of Lake Erie. A beach that had been losing sand was renourished in 1960–1961 by dumping 525,000 m^3 of sand, but this was soon depleted because the fill had a higher proportion of fine sand than the preceding natural beach. When the beach was renourished in 1965 with 12,700 m^3 of coarser sand it became more stable (Berg and Duane 1968).

Demands for the restoration of Hampton beach in Victoria, Australia, resulted in 1975 in the pumping of sand from the floor of nearby Sandringham Harbour through a pipeline on to the shore. The renourished beach was washed away by storm waves within a few days, because the sand taken from the harbour floor sand was much finer than that on the natural beach. This was because the sand that had drifted into the harbour had previously been sorted by wave action, which had left coarse sand on the beach and withdrawn the finer fraction to the sea floor, whence it was transported by nearshore currents into Sandringham Harbour (Bird 1996).

Several other beach renourishment projects have shown that if the sand used in beach renourishment is too fine it will quickly wash away. At Wrightsville, North Carolina, sand dredged from a nearby estuary was used to renourish the beach five times between 1939 and 1970. In each case the sand was removed during storms and deposited offshore (Pearson and Riggs 1981), and it was realised that the sand taken from the estuary was too fine for long-term retention. Renourishment with coarser sediment in 1972 was more successful.

On the island of Norderney the authorities renourished the beach in 1951 using sand dredged from nearshore shoals. The dredged sediment included a higher proportion of fine sand than was present on the natural beach, and this was soon extracted by wave action and washed back out to the nearshore zone (Eitner and Ragutzki 1994). Later renourishment projects have taken account of the fact that this sorting would occur, and that only the coarser fraction would persist on the beach, the finer material moving out to maintain nearshore bar formations (Kunz 1990).

Selection of grain size in sediment used for beach renourishment should also take account of wave energy conditions. This was illustrated on the often stormy shores of the Gulf of Georgia near Vancouver, Canada. Sand was introduced in an attempt to form a beach that would stop the erosion of cliffs cut in unconsolidated glacial deposits, but it was soon dispersed by storm waves. A second project, using cobbles, was more successful in establishing a beach to protect the cliffs (Downie and Saaitink 1983). In general, coarser sediment is needed to stabilise a beach in a high wave energy environment, but fine sand may be retained on a low wave energy shore. At Nunn's Beach, adjacent to Portland Harbour, Victoria, Australia, fine sand that had accumulated south of the harbour breakwater was dredged and trucked round to widen the beach in 1990. In this sheltered environment the renourished beach of fine sand has persisted.

Beach renourishment using sediment larger than the natural beach sediment (often termed 'beach coarsening') can be carried out, with the deliberate aim of creating a beach more resistant to erosion. Beach coarsening is normally restricted to beaches that have been severely eroded and are failing to give an adequate level of backshore protection (CIRIA 2010).

The durability of a beach may depend on the shape of the sand or pebbles used, angular material being less readily transported and lost than well-rounded material. In the 1960s an experiment was conducted in St Lucie County, Florida, when 1,000 tons of imported oolitic sand from the Bahama Banks and a similar quantity of local beach sand were placed in rows across the shore. These were redistributed by wave action during high tides, and it was found that the more angular oolitic sand was less mobile than the well-rounded native sand, and therefore more likely to persist on a renourished beach (Cunningham 1966).

It is necessary also to take account of rates of weathering of beach sediment, which can reduce grain size. On Delray Beach in Florida it was found that organic sand dredged from the nearshore zone was brittle, and when placed on the beach it disintegrated rapidly under wave agitation to finer sediment that was quickly dispersed. The specific gravity of introduced sediment is another relevant factor. In Tauranga Harbour, New Zealand, a renourished beach contained pumice, but this very light sediment was soon washed away, indicating that wave action selectively removes lighter as well as finer sediment (De Lange and Healy 1990).

Where gravel dredged from the sea floor contains sand and silt its use in beach renourishment can result in the formation of an excessively compact and impermeable beach capping, which may develop a firm cliff near the high tide line. Examples of such scarping have been documented from renourished beaches at Whitstable and Hayling Island in south-east England and Lodmoor near Weymouth in Dorset. It is regarded as a hazard to beach users, and a cause of reflective scour by waves on the lower beach during storms (McFarland et al. 1994).

Suitability of material available for beach fill was a problem at Aberystwyth, where in 1963 1,530 m^3 of waste from a quarry on Constitution Hill, to the north, was dumped on the shore to improve the beach at Victoria Terrace. It contained a high proportion of laminated and soft shale, which disintegrated and was quickly dispersed (So 1974).

4.4.3 Cross-Shore Profile and Determination of Volume

Similar considerations apply to the shape in profile. The profile of a beach renourishment is a geometric shape typically consisting of a dry beach berm and a foreshore slope. Cost constraints, environmental issues and sponsor preferences commonly influence the steepness of the slope and the length of the dry berm (USACE 2002), both determining total volume. As discussed earlier, profile design can improve beach stability and reduce the amount of adjustment following renourishment. In the Dutch method (Sect. 4.4.1, p. 79) the volume of sediment required is calculated using data on historical rates of erosion, and the sediment is then placed between −1 m below low tide line and the dune foot on the sandy beaches.

When considering the renourishment of the whole profile, the aim is to establish a relatively stable 'equilibrium profile'. Beaches with concave profiles are more stable than those with straight, convex or irregular profiles, and once concave profiles are attained they become relatively (but not absolutely) stable. The gradient of the concave profiles varies with grain size and preceding wave conditions. Subsequent oscillations occur with episodes of storm wave erosion and fine weather accretion: the profile following storm wave erosion is different from a calm weather profile. As long as the beach profile oscillates between these limits the beach can be considered relatively stable, but as has been noted (Sect. 1, p. 2) beach erosion is prevalent and most renourished beaches need further inputs of beach sediment (Sect. 4.5, p. 84).

In the United States much attention has been given to theoretical equilibrium profiles (Campbell and Benedet 2006). It is thought that beaches tend towards an equilibrium profile and that a newly renourished beach will adjust over time to achieve this equilibrium (Dean 1977, 1991). An equilibrium beach slope can be predicted, given the grain size of renourishment sediment; conversely, an appropriate grain size can be chosen to provide a required equilibrium beach slope. The latter was demonstrated by Firman et al. (2011) where a particular profile was required to avoid renourishment sediment encroaching on adjacent offshore seagrass habitat. At Colwyn Bay in the United Kingdom, grain size was chosen to create the most stable beach profile and beach width in response to storms (Oliveira et al. 2011). Testing of a number of options demonstrated that using sediment with a mean grain size diameter of 0.45 mm was more stable than a mean grain size diameter of 0.25 mm.

Depth of closure is defined as the depth (and so distance offshore) beyond which no significant profile fluctuation takes place due to coastal processes (wave and current action) (Hallermeier 1981). A renourished beach can extend out to the depth of closure, but beyond this sediment deposited will not contribute to the maintenance of the beach. This concept is of little use on oceanic coasts: on the New South Wales coast the depth of closure is typically 22 m (at about 1.2 km offshore) (Neilson 1994), but at Salina Bay in Malta the depth of closure was estimated to be 5 m (approximately 100 m offshore), and was used to limit the movement of renourished sediment on to sea floor seagrass habitats, which occurred during brief periods of storm wave action (Firman et al. 2011).

A renourished beach berm should generally correspond to the level of the natural berm crest, determined from historical conditions at the site (USACE 2002). If this is not possible the natural berm height should be estimated from survey data from berm levels on nearby beaches exposed to similar waves. Beach width depends on the project objectives. For example, where a beach is being renourished to improve amenity, the beach may need to maintain a minimum width, while a minimum beach width following storms may be required to reduce damage to adjacent property.

4.4.4 Beach Configuration and Orientation

There have been suggestions that once a beach attains a particular shape in plan it will become stable. For example, beaches shaped by obliquely-arriving waves alongside a headland or breakwater develop an asymmetrical curvature sometimes known as a crenulate, 'half-heart' or 'zeta-curve' configuration. The notion that 'headland breakwaters' can be used to shape stable renourished beaches within intervening compartments by attaining such a configuration, related to the refraction of obliquely arriving waves, is based on the work of Silvester (1976), who indicated that beach stability (in the sense of zero longshore drift) could be attained when bays assumed this configuration. Headland breakwaters were constructed on the shore of East Coast Park in Singapore, where it was found that longshore drift diminished as the intervening beaches attained a crenulate shape, but erosion continued on the asymmetrical beaches formed in this way, so that the problem of coastline instability remained.

Nevertheless, some beach outlines are more stable than others, and a renourished beach will be more persistent if it is placed on an alignment that is compatible with incident wave regimes. Reference has been made to the importance of aspect in relation to prevailing wave patterns in determining directions of longshore drift (Fig. 4.21), notably on the north-east coast of Port Phillip Bay, Australia. The renourished beach at Wrightsville, North Carolina was initially incorrectly aligned, but became more stable as the beach became realigned more closely to the pattern of incoming waves. Breakwaters or terminal groynes can be used to delimit a renourished beach in such a way that it would become correctly aligned. Attention to dominant wave regimes and patterns of wave refraction approaching a coastal sector can improve a beach renourishment project by selecting a suitable alignment for the project design. At Mission Bay and Kohimarama in Auckland, New Zealand, renourished beaches contained between groynes were built with an orientation related to the incident wave energy. They remained relatively stable (Papps and Priestley 2005). On the coast of Malta an artificial beach was orientated to fit the pattern of incident waves (Firman et al. 2011).

4.5 Monitoring Changes After Beach Renourishment

Changes will occur on beaches that have been renourished. These beaches will be eroded by the same processes that depleted the preceding natural beaches (Riddell and Young 1992). It is often observed in practice that erosion, sediment volume loss and coastline retreat, are greater on a renourished beach than the historical rates on a natural beach (Dean 2000). As a result it is impossible to meet the expectations of stakeholders who expect renourishment to last indefinitely. Most renourished beaches begin to lose sediment as soon as they have been emplaced, some of the beach material being washed or blown away alongshore, some being swept to the backshore and beyond, and some withdrawn to the sea floor.

Monitoring of coastal processes and morphology following beach nourishment is necessary to gain understanding of the underlying causes of beach erosion, and improve subsequent project design (NRC 1995), while providing guidance and calibration for numerical models (Dean 2002). Monitoring and mapping are also necessary to determine the rates and patterns of sediment losses, indicating when and how much supplementary beach material is required and where it should be placed (Foxley and Shave 1983). Changes are usually measured by making repeated surveys along transverse profiles from the back of the renourished beach out on to the nearshore sea floor, and linking these by alongshore surveys. The use of remote sensing techniques is also common, such as LIDAR survey and satellite imagery. Supporting evidence is usually available from sequential air photographs and video imagery has also been used (Elko et al. 2005, Ojeda et al. 2008). In the absence of sufficient profile survey data or wide-coverage satellite imagery Raman Murthy et al. (2008) used remote sensing to monitor coastline changes following a beach renourishment project on the southeast coast of India. Gares et al. (2006) examined the use of LIDAR data for beach renourishment monitoring at Wrightsville Beach, North Carolina, and demonstrated that this information provides data for use in both horizontal (shoreline) and volumetric (sediment budget) analyses of changes following renourishment.

Multiple sources of data can be useful when monitoring changes following a renourishment. Since the completion of the 'sand engine' project on the Dutch coast (Sect. 4.3.2, p. 62, 63) in 2011, the topographic evolution of the renourished profile has been monitored monthly using a purpose-built jet-ski mounted with a global positioning system and an echo- sounder, in addition to four-yearly coastal profile measurements (Stive et al. 2013). Furthermore, two high-resolution video cameras have been installed overviewing the 'sand engine' and adjacent beaches, and regular aerial photographs are collected.

The aim of monitoring is to understand the processes that erode and distribute emplaced beach material. Such information can guide future beach management, such as the insertion of groynes, the introduction of regular renourishment updrift on beaches that are losing sediment alongshore, localised renourishment at places of severe erosion, or the need to repeatedly restore the profile of a beach that is losing sediment offshore.

Long-term monitoring of several renourished beaches on the Atlantic coast of the United States showed rapid initial losses followed by more gradual depletion. At Long Beach Island, New Jersey, for example, the beach renourished in 1979 eroded rapidly for the first eighteen months, then more slowly, until by 1986 all of the sand that had been added had disappeared, and the beach had returned to its pre-1979 width and profile. Rapid initial losses were also observed at Upham Beach, Florida by Elko et al. (2005). Following nourishment in 1996 high resolution video images were collected concurrent with surveyed beach profiles to investigate planform evolution. 193,000 m^3 of sediment placed across 0.6 km of shore advanced the coastline by up to 175 m. Observations demonstrated that erosion then occurred at a rate of approximately 70 m/year in the first year and 135 m/year in the second year, principally by losses downdrift. 50 % of the placed beach material was removed within a year, and by 2000 (4 years after placement) all of the placed beach sediment had disappeared.

As changes proceed the plan and profile of the beach are re-shaped into patterns more closely adjusted to the prevailing wave and current regimes, influenced by the grain size of the remaining sediment (Everts et al. 1974; Blott and Pye 2004): usually the finer sediment is removed first, leaving the beach coarser in texture, and often steeper in profile.

At Mentone on the shores of Port Phillip Bay, Australia, the beach terrace formed by renourishment in 1977 (Sect. 4.3.1, p. 55) was monitored by repeated surveys of transverse profiles at 30 m intervals. These showed a reduction of beach terrace width from 32 to 22 m in the first year, due to re-shaping of the seaward slope to a slightly concave profile with an average gradient of 1 in 9. At the same time, parallel bars of fine sand formed in the nearshore zone, partly from the finer fraction withdrawn from the restored beach material by storm waves, and partly from the fine silty sands already present on the sea floor. The beach terrace was thereafter cut back about a metre per year, recession occurring mainly during brief episodes of storm wave activity, especially when these coincided with high tides. In 1984 Mentone Beach was renourished by pumping in a further 18,500 m^3 of coarse sand to restore the profile to its 1977 dimensions (Jones and Schafter 1986). This has been effective, for although there has been further gradual depletion, the beach still had an average width of 20 m in 2014.

The value of monitoring a renourished beach was well illustrated at Virginia Beach, on the Atlantic coast of the United States. Numerous groynes had been built in the late 1940s in the hope of halting beach erosion, but these failed to retain the beach, and in 1952–1953 just over a million m^3 of sand, coarser than that on the original beach, was added. Monitoring showed continuing depletion of this beach, and it became clear that the groynes were of little use, because most of the sand was being withdrawn to the sea floor, instead of drifting away alongshore. It is necessary to determine the direction of beach losses before inserting groynes, which are more effective when beach material is moving alongshore rather than seaward. The conclusion was that Virginia Beach could only be maintained by frequent renourishment or backpassing of sand lost to the sea floor. This approach to the replenishment of Virginia Beach illustrates Pilkey's (1990) suggestion that beach renourishment projects should be regarded as experimental, with improvements based on experience gained from continued monitoring.

Mapping of changes on a renourished beach can determine patterns of movement alongshore. In 1963 sand was dumped on the shore near Absecon Inlet, north-east of Atlantic City, New Jersey. Repeated mapping showed that the beach fill was shaped into a lobe that migrated south-westward along the shore at 2–3 m/day.

This was a smaller version of the 'sand engine' technique described on Sect. 4.3.2, p. 62, 63. After 2 years this lobe arrived to augment the beach in the vicinity of the main pier, but it continued to move along the shore in front of the boardwalk, and then beyond, so that the widening of the Atlantic City beach was only temporary. The response in 1970 was to add a further 596,000 m^3 of sand near Absecon Inlet, and this also moved alongshore to the Atlantic City pier, where the beach was widened in 1972, and moved on south-westward (Everts et al. 1974). It was then clear that on this longshore drift-dominated coast a beach could only be maintained at Atlantic City by frequent small injections of sand at the north-eastern end (Pilkey and Clayton 1987). Groynes were placed on the shore to reduce the rate of drifting to the south-west in an attempt to keep the beach at Atlantic City (Weggel and Sorensen 1991).

Monitoring of renourished beaches has generally been restricted to the emplaced sector, to decide when and where further renourishment is necessary, but there should also be mapping and monitoring of changes on adjacent shore and nearshore areas to which eroded sediment may move. There is a risk that sediment from renourished beaches will drift alongshore and accumulate in boat harbours, or as shoals impeding navigation at the mouths of rivers and creeks. On the other hand, it may prove beneficial in renourishing other beaches along the coast.

Monitoring can determine quantities of sediment lost from a beach. At Kirra, at the southern end of the Gold Coast in Queensland, Australia, over 5 million m^3 of sand was dumped on the beach in several phases between 1985 and 1990. A survey in May 1992 showed that 87 % of this renourished sediment was still on this beach or in the nearshore region, the remainder having drifted alongshore to augment beaches to the north (Delft Hydraulics Laboratory 1992).

Decisions on when and where a renourished beach should be further replenished, and how much beach fill is required, can be made in terms of information from such mapping and monitoring. There were rapid changes after renourishment on the beach at Wrightsville, North Carolina, monitoring of 50 transverse profiles showing rapid initial losses, some 66 % of the renourished sediment being lost within the first year. The erosion rate slackened as the beach profile, originally a terrace with a convex seaward slope, became concave, and there was then more gradual recession, the beach maintaining a more or less constant seaward slope. The initial rate of erosion on the renourished beach was ten times that of the preceding natural beach erosion (Pilkey and Clayton 1987), but the loss rate on the renourished beach declined to the long-term natural erosion rate after 8 years. This was the effective residence time of the beach fill, and indicated that if the renourished beach were to be maintained here it would need to be renourished at intervals of about 5 years.

Management of the Adelaide beaches in South Australia also used monitoring of changes in beach width and profile to determine where and how much sand replenishment was needed (Fotheringham and Goodwins 1990). Instrumental surveys have been carried out since 1975 on a series of beach profiles spaced approximately 750 m apart, with closer monitoring where necessary at 50 m intervals. The results were presented in the form of maps that shade areas with surface gains or losses of between 0.2 and 1.0 m, and more than 1.0 m. More recently the data have been processed using Geographical Information Systems to produce coloured contour maps of the beaches, from which patterns of gain and loss can be identified, and areas of developing deficit replenished by dumping sand (Noyce 1993).

On the shores of the Great Lakes in North America monitoring has shown that there have been changes on renourished beaches accompanying irregular oscillations in water level of up to 5 m over periods of several years. When lake levels rise beach erosion occurs, and when they fall there is progradation. In 1974 175,000 m^3 of beach fill was emplaced on the shore of Michigan City, Lake Michigan, and a further 61,000 m^3 in 1981. The emplaced beach profile was soon modified by wave action, becoming relatively stable in relation to the variable lake levels (Jansen 1985). Renourished beach profiles thus adapt to hydrodynamic variations, often with a time lag of several weeks or months (Thompson 1987). The United States Army Corps of Engineers has since renourished beaches at several other sites on the Lake Michigan coast, using coarse sand dredged from the lake floor below the 5 m contour. Most of these beaches have remained in position even during phases of high lake level (Macintosh and Anglin 1988).

4.6 Assessment of Beach Performance

Changes on renourished beaches are often rapid, and there is disappointment and criticism when an emplaced beach quickly diminishes. A completed beach renourishment project typically lasts between 3 and 10 years depending on the site, project plan, and number and intensity of storms (Weggel 1995): this is referred to as beach durability. The time between placement and loss of 50 % of the renourishment volume has been termed the half-life (Leonard et al. 1990b) and can be used as a measure of beach durability when assessing the performance of a beach renourishment (Elko et al. 2005; USACE 2002).

There has been much discussion of beach renourishment performance, particularly on the coasts of the United States. Before the 1950s beach renourishment projects on the Atlantic coast were intuitive, without much planning or design, and there seemed to be an assumption that the sandy beaches between New York and Miami were all more or less the same. Subsequently more attention was given to scientific research, acknowledging that there are variations in beach morphology, aspect, and nearshore conditions as well as contrasts related to the location and dynamics of tidal inlets. Most Atlantic coast beaches are of sand washed in from the sea floor, but there are some fluvially-fed sectors and in the north some areas of cliff-derived beach sediment.

Important parameters in evaluating the performance of a renourishment often include the dry beach width, the volume of sand remaining after a storm, and the remaining subaqueous sand volume (NRC 1995). Coastline location is also used as a key performance indicator, as in Holland where maintaining the 1990 coastline is one objective, while in the UK beach renourishment performance is often measured against the standard of protection offered against flooding.

Controlling factors of nourishment performance vary from one project to another, as well as spatially and temporally. For example, in three phases of renourishment at Sand Key Beach in west central Florida, controlling factors on performance included location in the regional sediment transport regime, magnitude of wave energy, sediment characteristics of the renourishment beach material, local reversals in longshore transport, presence of hard structures, adjacent beach nourishment, variation in coastline orientation, and beach fill technique (Davis et al. 2000).

Nevertheless, there are still doubts about the durability of beach renourishment projects. Walton and Purpura (1977) found that several renourished beaches on the Atlantic coast had performed poorly, and this they attributed to the widespread use of undersized material, renourishment too close to tidal inlets, and unexpectedly frequent storm activity. The proximity of beach renourishments to tidal inlets was also considered a key factor in renourishment performance by Roberts and Wang (2012). They used post-renourishment monitoring of a number of barrier island beaches renourished in 2006, including Sand Key, Treasure Island and Long Key in west-central Florida to demonstrate the importance of tidal inlet processes on renourishment performance.

Pilkey and Clayton (1987, 1989) critically reviewed more than 90 beach renourishment projects on the Atlantic coast and found that few had persisted as long as originally predicted while most of them had proved far more costly than anticipated. South of Cape Kennedy engineers had been more successful in predicting the fate of replenished beaches, with Miami Beach a notable success, but on most of the beaches on the Atlantic coast the sand deposited had been completely washed away in less than 5 years (26 % in less than a year), usually because of erosion during storms; only 12 % had persisted for more than 5 years (Leonard et al. 1990a, b). Moreover, the renourished beaches had not recovered from hurricanes as quickly as natural beaches.

In an editorial in the *Journal of Coastal Research* in 1990 Pilkey noted that storms seemed to have been the major factor determining renourished beach longevity on the Atlantic coast, unpredicted erosion often being attributed to unusual storm activity. The public were told that a replenished beach would recover during fair weather, that loss rates would diminish over time, and that the lost sand had moved offshore and would diminish wave energy on the depleted beach, so that the next beach renourishment would last longer.

Questioning the success of Atlantic coast beach renourishment projects led to a spirited discussion in the *Journal of Coastal Research* by Houston (1990), Pilkey and Leonard (1990) and Houston (1991). This indicated that documentation of beach changes, both before and after renourishment, had been inadequate, that there was a need for the public to be better informed on how renourished beaches may perform, and a need to support accurate and sustained monitoring. A U.S. Army Corps of Engineers (1994) report examined more than 100 replenished beaches and concluded that actual costs and volumes of sand placed were within 5 % of predicted values (Houston 1995; Sudar et al. 1995), but Pilkey (1995) cited some omitted problems, including the fact that some beaches were severely depleted between renourishments. At Tybee Beach, Georgia, where the costs and volumes of beach renourishment in 1976 and 1987 were indeed less than predicted, the first placement disappeared within a year, so that for a decade Tybee was without any beach. Damage done to backshore property and structures should really be included as a cost item. Predictions remained uncertain: the 1993 renourishment of Folly Beach, South Carolina, was expected to require repetition every 8 years, but this was already necessary after 1 year (Pilkey 1995).

One response to the need to retain renourished sediment on the shore is seen at Porthcawl in South Wales, where tarmac has been introduced to stabilise a shingle beach (Fig. 4.30).

It is now generally acknowledged that renourished beaches will be eroded, and will have to be replaced at intervals, and that this will require substantial and ongoing expenditure by governments and coastal communities. The alternatives are to use solid protective structures, which do not co-exist well with beaches, or to allow natural changes to proceed on the coastline, abandoning eroding land (the process termed managed retreat). It seems likely that demands for beach renourishment will continue as a component of comprehensive coastal management programmes, because of increasing coastal population and development stimulating further demands for beach recreation, because of greater public awareness of beach erosion problems and because of widespread opposition to the use of hard structures in coastal protection. Objections to nearshore dredging and truck traffic as means of obtaining and transporting sediment for beach renourishment were voiced in Adelaide, South Australia (Sect. 4.3.4, p. 67), but are likely to fade as

Fig. 4.30 The shingle beach at Porthcawl, South Wales, stabilised in tarmac. © Geostudies

demands for beach renourishment intensify. On the barrier islands of the Atlantic coast of the United States local residents now regard truck traffic as acceptable if it is the only way of maintaining their beaches.

Public demands for beach renourishment are often a response to obvious depletion by storms, and the fear that further losses would ensue. At Long Beach Island, New Jersey, such demands led to renourishment of a depleted beach in 1979, but the added sediment had disappeared after 7 years. There is a divergence of longshore drift here, sand from the northern part of the beach being washed back into Barnegat Inlet (from which it had been obtained) while sand from the southern part moved away southward along the coast (Ashley et al. 1987). Although narrow, this renourished beach became relatively stable, except for brief and temporary depletions during stormy periods, and it is now locally regarded as acceptable. Public perception of erosion hazards can be fickle: in the words of Pilkey and Clayton (1987) "community apprehension over a narrow beach softens with time and the absence of storms". There was also the feeling that the beach could, if necessary, be restored again, this time with sand deposition concentrated in the zone of divergence, and losses northward and southward perhaps delayed by the insertion of groynes.

It has been noted (Sect. 4.3.1, p. 59 and Sect. 4.4.2, p. 79) that beach renourishment should use sediment at least as coarse as that on pre-existing natural beaches, because finer sediment is quickly lost. Coarse beaches are less attractive to beach users (Campbell and Beachler 1984), although more durable shingle beaches are used for recreation at seaside resorts such as Brighton in southern England, Dieppe in northern France and Nice on the French Riviera.

4.7 Modelling of Beach Renourishment

More detailed modelling (both physical and numerical) may be needed to refine the design of a renourishment project, or to investigate how it will alter following placement.

Laboratory simulation of shore processes to help design beach renourishment projects has been attempted with scale models such as water tanks in which waves, tides and currents can be generated and their combined effects assessed. The aim is to test hypotheses concerning the ways in which these processes cause erosion, move sediment and promote deposition on the sea floor and along the coast. Such physical models have limitations because of the difficulty of scaling down materials and processes without modifying their physical properties (e.g. coherence, friability, expansion and contraction of sediments; viscosity and surface tension in water), but they have been useful in exploring potential responses to marine and nearshore processes (Silvester 1974). They are also useful in examining the impacts of structures, as at Borth in Wales (Fig. 4.31), where the future evolution of the beach following renourishment was examined in relation to the introduction of groynes and a submerged breakwater.

Numerical modelling has been much used by engineers as a basis for computer simulations of coastal processes (hydrodynamics and sediment transport),

Fig. 4.31 Physical modelling being undertaken to inform coastal protection works in Borth, UK, comprising beach renourishment, offshore breakwaters and submerged breakwaters.
© Mick Newman (of Royal HaskoningDHV)

especially since 1970 (Chou et al. 1983). Such modelling can be used to study the effects of integrated processes (waves, tides and currents) on nearshore sediment flow, and the ways in which these processes and responses will be modified by the introduction of structures such as groynes or by beach renourishment. It is important to be sure that the information used is accurate and comprehensive, and to test predictions against what actually happens, in order to refine the model and improve subsequent forecasts.

Numerical modelling is useful as a means of exploring process–response relationships, but as coastal systems are complex and predictions can prove unreliable. Monitoring of coastal changes is needed to check predictions and obtain further data for refinement of models. Numerical models can also be used to assess impacts and implications, and interpret field data.

It is necessary to consider the capabilities and limitations of each model. Numerical models are sensitive to morphological as well as input (forcing) parameters and so must be calibrated against measured data (Hamm et al. 2002). When validated, numerical modelling can be used to examine several alternative beach nourishment strategies and optimise their performance (Hanson et al. 2002; Capobianco et al. 2002). Nevertheless, the complexity of coastal and beach processes requires that complete reliance on numerical modelling should be avoided and predictions of beach renourishment performance should be firmly grounded on relevant experience and expertise. The use of numerical modelling for all aspects of beach renourishment has been questioned by several authors, including Cooper and Pilkey (2004), who advocate the use of a conceptual approach as an alternative to numerical modelling of such a complex system. The authors suggest seven conceptual alternatives, including past engineering experience on the beach in question or on neighbouring beaches, global experience on similar beaches, the use of geo-indicators and field studies.

Numerical models were used in the design of a beach renourishment project at Salina Bay, Malta (Sect. 4.4.3, p. 82), where wave patterns and responses were examined, and the results used to determine an optimum beach configuration for long-term stability (Firman et al. 2011).

At West Beidaihe Beach, south of Qinhuangdao in China, removal of a rock jetty in 2002 for safety and aesthetic reasons resulted in the retreat of the coastline and the reduction of a renourished beach (Kuang et al. 2011). Kuang et al. (2011) used numerical models to find that, without further renourishment, West Beidaihe Beach would cease to provide a subaerial beach area suitable for recreation within 2–3 years. After numerical evaluation of various options it was decided that a groyne at the eastern end of the beach played a key role in retaining the beach in the long term. More sand was placed on the beach, and protected by a 250 m long submerged breakwater.

At Anna Maria Key, Florida project design of the renourishment project completed in 1993 used modelling and successfully estimated a nine year renourishment interval (Dean 2002). At Colwyn Bay Beach, Wales numerical modelling was used to assess alternative options for beach protection. The assessment was divided in three phases: assessment of beach dynamics, definition of an optimum recharged beach profile, and long-term modelling of alternative solutions (Oliveira et al. 2011). The large quantity of data provided by the assessment of the beach dynamics was used as input and verification data for the numerical models, which were then applied to test factors such as profile response to storms for different berm widths and sediment sizes and to identify an optimum renourishment beach profile.

Predictions of the direction and quantity of longshore drift formed part of the design and environmental approval process for the 'sand engine' model (Sect. 4.3.2, p. 62) (Stive et al. 2013). Morphodynamic numerical modelling was used to obtain projections of the temporal and spatial evolution of the large-scale renourishment, 3, 5, 10, 15 and 20 years after renourishment.

Models have been used in the United States to estimate the impacts of dredge site (borrow pits) on adjacent beaches (Benedet et al. 2013), providing valuable information about how different dredge pit designs interact with the adjacent shore. With such information adjustments to the design of these features can be made to minimise impacts on the adjacent coast.

Hartog et al. (2008) used numerical modelling, consisting of analyses of waves, hydrodynamics, and a morphology to identify the processes that influenced the performance of renourishment at Delray Beach, Florida. The analyses included an assessment of the impacts of borrow pit location and size on the renourishment, changes alongshore, and erosion due to the change in coastline orientation caused by deposition of the renourishment.

In addition to numerical modeling and laboratory-based assessment, field experiments have been used to help design and assess performance of beach renourishments. One such field assessment, sediment tracing, is a useful method for quantifying the magnitude and direction of littoral sediment flux. Provided the tracer material faithfully represents the native sediment characteristics (e.g. in particle size and fall velocity), tracer can indicate littoral transport through monitoring over time and space (McComb and Black 2004). As well as use in determining the fate of dredge material (Marsh et al. 1997) and sediment transport associated with by-pass systems (Sherman et al. 1990, Uda et al. 1991), sediment tracing has been used in the design of a beach renourishment project by developing a better understanding of the sediment transport processes (Fig. 4.32). Sediment tracing can also be used to evaluate the performance, with tracing material incorporated into the beach renourishment sediment.

At the Great Egg Harbour inlet in New Jersey on the east coast of the United States, sediment tracer studies were conducted to understand the fate of beach renourishment sediment. Tracer material was placed at the low water line on the beach and then the movement of the tracer monitored over a 30-day period through the collection and analysis of sediment cores up and down the beach, as well as on the inlet ebb shoal.

Fig. 4.32 **a** Indicator sediment (tracer) being released at the start of a study in 2014. **b** Example of a recovered core including tracer material. © Royal HaskoningDHV/Jon Marsh of ETS Worldwide

At the mouth of the Columbia River, studies have been carried out on behalf of the Portland District of the US Army Corps of Engineers to understand the beneficial re-use of dredge material for beach renourishment (Moritz et al. 2011). Sediment tracers were used to estimate the dominant sediment pathways of dredge material placed near the mouth of the Columbia River, where the sediment was expected to be transported to an adjacent beach.

References

Acworth C, Lawson S (2011) The Tweed River entrance sand bypassing project ten years of managing operations in a highly variable coastal system. In: Proceedings of the 20th NSW coastal conference, Tweed Heads, pp 1–23

Anders FJ, Hansen M (1990) Beach and borrow site investigation for a beach nourishment at Ocean City, Maryland. US Army Corps of Engineers, Waterways Experiment Station, Coastal Engineering Research Center, Technical Report 90–5

Ashley GM, Halsey SD, Farrell SC (1987) A study of beach fill longevity: Long Beach Island, New Jersey. Proc Coast Sediments 87:1188–1202

Benedet L, Finkl CW, Dobrochinski JPH (2013) Optimization of nearshore dredge pit design to reduce impacts on adjacent beaches. J Coastal Res 29:519–525

Berg DW, Duane DB (1968) Effect of particle size and distribution on stability of artificially filled beach, Presque Isle Peninsula, Pennsylvania. In: Proceedings 11th conference Great Lakes research, pp 161–178

Bird ECF (1991) Changes on artificial beaches in Port Phillip Bay, Australia. Shore Beach 59:19–27

Bird ECF (1996) Beach management. Wiley, Chichester

Blott SJ, Pye K (2004) Morphological and sedimentological changes on an artificially nourished beach, Lincolnshire, UK. J Coastal Res 20:214–233

Boak EH, Jackson LA, McGrath JE, Brosnan MP (2001) An overview of gold coast coastal management, 1960–2001. In: Coasts and ports conference 2001: proceedings of the 15th Australasian coastal and ocean engineering conference, the 8th Australasian port and harbour conference, Institute of Engineers, Australia, pp 561–566

Boswood PK, Murray RJ (2001) Worldwide sand bypassing systems: data report. Coastal Services technical report R20. Conservation technical report No. 15 (online). Report produced for the Queensland Government, Environmental Protection Agency. 2001. ISSN 1037-4701

Boswood P, Voisey CJ, Victory SJ, Robinson DA, Dyson AR, Lawson SR (2005) Beach response to Tweed River entrance sand bypassing operations. In: Proceedings of the 2005 coasts and ports conference, Australia, pp 124–199

Bruun P (1990) Beach nourishment—improved economy through better profiling and backpassing from offshore sources. J Coastal Res 6:265–277

Bruun P (1996) Navigation and sand bypassing procedures at inlets: technical management and cost aspects. J Coastal Res 23:113–119

Bruun P, Willekes G (1992) Bypassing and backpassing at harbours, navigation channels and tidal entrances: use of shallow-draft hopper dredgers with pump-out capabilities. J Coastal Res 4:687–701

Campbell TJ, Beachler K (1984) The 1983 beach restoration of Pompano and Lauderdale by the sea. Shore Beach 52:21–26

Campbell TJ, Benedet L (2006) Beach nourishment magnitudes and trends in the U.S. J Coastal Res 39:57–64 (special issue)

Campbell JF, Moberley R (1984) USA-Hawaii. In: Walker HJ (ed) Artificial Structures and Shorelines. Louisiana State University, USA, pp 1–4

Capobianco M, Hanson H, Larson M, Steetzel H, Stive MJF, Chatelus Y, Hamm L, Aarninkhof S, Karambas T (2002) Nourishment design and evaluation. Applicability Model Concepts 47:113–135

Castelle B, Lazarow N, Marty G, Tomlinson R (2006) Impact of beach nourishment on Coolangatta Bay morphology over the period 1995–2005. In: Proceedings of the 15th NSW coastal conference, pp 1–12

Chill J, Butcher C, Dyson W (1989) Beach nourishment with fine sand at Carslbad, California. Proc Coast Zone 89:2091–2103

Chou IB, Powell GM, Winton TC (1983) Assessment of beach fill performance by excursion analysis. Proc Coast Zone 83:2361–2386

Ciprani LE, Dreoni AM, Pranzini E (1992) Nearshore morphological and sedimentological evolution induced by beach restoration: a case study. Bollettino di Oceanologica Teorica ed Applicata 10:279–295

CIRIA (2010) Beach management manual (report no. C685). CIRIA, London

Clarke J, Brookes S (2008) Practical aspects of executing renourishment schemes on mixed beaches. Science Report SCH00608B0FD-E-P (online), Environment Agency

Cooper NJ (1998) Assessment and prediction of Poole Bay (UK) sand replenishment schemes: application of data to Fuhrboter and Verhargen models. J Coastal Res 14:353–359

Cooper JAG, Pilkey OH (2004) Alternatives to the mathematical modelling of beaches. J Coastal Res 20:641–644

Coughlan PM (1989) Noosa Beach—coastal engineering works to mitigate the erosion problem. In: 9th annual conference coastal & ocean engineering, pp 198–203

Cunningham RT (1966) Evaluation of Bahamaian oolitic aragonite sand for Florida beach nourishment. Shore Beach 34:18–21

Davis RA Jr, Wang P, Silverman BR (2000) Comparison of the performance of three adjacent and differently constructed beach nourishment projects on the Gulf Peninsula of Florida. J Coastal Res 16:396–407

De Lange WP, Healy TR (1990) Renourishment of a flood-tidal delta adjacent beach, Tauranga Harbour, New Zealand. J Coastal Res 6:627–640

Dean RG (1977) Equilibrium beach profiles: U.S. Atlantic and Gulf Coasts. Ocean Engineering Report 12

Dean RG (1983) Principles of beach nourishment. In: Komar PD (ed) Handbook of coastal processes and erosion. CRC Press, Boca Raton, pp 53–84

Dean RG (1991) Equilibrium beach profiles: characteristics and applications. J Coastal Res 7:53–84

Dean RG (1995) Beach renourishment planform considerations. In: Proceedings, coastal dynamics 1995, American Society of Civil Engineers, pp 533–546

Dean RG (2000) Beach renourishment design: consideration of sediment characteristics. In: Proceedings of the 2000 national conference on beach preservation technology. Florida Shore and Beach Preservation Association, pp 41–63

Dean RG (2002) Beach nourishment: theory and practice. Advanced series on ocean engineering, vol 18, World Scientific Publishing, p 399

Delft Hydraulics Laboratory (1992) Southern Gold Coast littoral sand supply. Delft, The Netherlands

Dornhelm RB (1995) The Coney Island public beach and boardwalk improvement of 1923. Shore Beach 63:7–11

Downie KA, Saaitink H (1983) An artificial cobble beach for erosion control. Proc Coast Struct 83:846–859

Duvivier P (1991) Mablethorpe to Skegness Sea Defences Strategy Study. Report to the National Rivers Authority. Peterborugh, UK

Eddison J (1983) The evolution of the barrier beaches between Fairlight and Hythe. Geogr J 149:39–53

Edge BL, Cruz-Castro O, Magoon OT (2002) Recycled glass for beach nourishment. In: Proceedings 28th international conference on coastal engineering. Cardiff, Wales, pp 3630–3641

Eitner V, Ragutzki G (1994) Effects of artificial beach nourishment on nearshore sediment distribution (Island of Nordeney, Southern North Sea). J Coastal Res 10:637–650
Elko NA, Holman RA, Gelfenbaum G (2005) Quantifying the rapid evolution of a nourishment project with video imagery. J Coastal Res 21:633–645
Evangelista S, La Monica GB, Landini B (1992) Artificial beach nourishment using fine crushed limestone gravel at Terracina (Latium, Italy). Bollettino di Oceanologica Teorica ed Applicata 10:273–278
Everard CE (1962) Mining and shoreline evolution near St. Austell Cornwall. Trans Roy Geol Soc Cornwall 19:199–219
Everts CH, De Wall A, Czerniak MT (1974) Behaviour of beach fill at Atlantic City, New Jersey. In: Proceedings of 14th conference on coastal engineering, pp 1370–1388
Finkl CW (1981) Beach nourishment, a practical method of erosion control. Geo-Mar Lett 1:155–161
Finkl CW (1993) Pre-emptive strategies for enhanced sand bypassing and beach replenishment activities in southeast Florida: a geological perspective. J Coastal Res 18: 59–89 (special issue)
Finkl CW (1996) Beach fill from recycled glass: a new technology for mitigation of localized erosional 'hot spots' in Florida. In: Proceedings of the 9th national conference on beach preservation technology (Tallahassee, Florida), Tallahassee, Florida Shore and Beach Preservation Association, pp 174–175
Finkl CW, Walker HJ (2004) Beach nourishment. In: Schwartz M (ed) The encyclopedia of coastal science. Kluwer, Dordrecht, pp 147–161
Firman K, Kemp J, Finch D, Mallia A, Sciortino J (2011) Designing a sustainable beach replenishment scheme for a site in Malta. J Coastal Res 61:36–43
Fotheringham DG, Goodwins DR (1990) Monitoring the Adelaide beach system. In: Proceedings of 1990 workshop on coastal zone management, Yeppoon, Queensland, pp 118–132
Foxley JC, Shave K (1983) Beach monitoring and shingle recharge. In: Shoreline protection conference, Telford, London, pp 163–170
Fulford ET, Grosskopf WG (1989) Storm protection project design, Ocean City, Maryland. Beach Preserv Technol 89:239–248
Galster RW, Schwartz ML (1990) Ediz Hook—a case study of coastal erosion and rehabilitation. J Coastal Res 6:103–114 (special issue)
Gares PA, Wang Y, White SA (2006) Using LIDAR to monitor a beach nourishment project at Wrightsville Beach, North Carolina, USA. J Coastal Res 22:1206–1219
Griggs GB (1990) Littoral drift impoundment and beach nourishment in northern Monterey Bay, California. J Coastal Res 6:115–126 (special issue)
Grove RS, Sonu CJ, Dykstra DH (1987) Fate of massive sediment injection on a smooth shoreline at San Onofre, California. In: Proceedings of coastal sediments, vol 87, pp 531–538
Guerin B (1984) Beach restoration, Mentone, Port Phillip Bay, Victoria. In: Knight MJ, Minty EJ, Smith RB (eds) Case studies in engineering geology, hydrogeology and environmental geology, vol 11. Geological Society of Australia, pp 398–410 (special publication)
Hall JV (1952) Artificially constructed and nourished beaches in coastal engineering. In: Proceedings of 3rd coastal engineering conference, pp 119–133
Hallermeier RJ (1981) A profile zonation for seasonal sand beaches from wave climate. Coast Eng 4:253–277
Hamm L, Capobianco M, Dette HH, Lechuga A, Spanoff R, Stive MJF (2002) A summary of European experience with shore nourishment. Coastal Eng 47:237–264
Hanson H, Brampton A, Capobianco M, Dette HH, Hamm L, Laustrup C, Lechuga A, Spanhoff R (2002) Beach nourishment projects, practices, and objectives—a European overview. Coast Eng 47:81–111
Hartog WM, Benedet L, Walstra DJR, Van Koningsveld M, Stive MJF, Finkl CW (2008) Mechanisms that influence the performance of beach nourishment; a case study in Delray Beach, Florida, USA. J Coastal Res 1304–1319
Healy TR, Kirk RM, De Lange WP (1990) Beach renourishment in New Zealand. J Coastal Res 6:77–90 (special issue)

Houston JR (1990) Discussion of editorial by O.H. Pilkey (1990) and paper by Leonard et al. (1990). J Coastal Res 6:1023–1035
Houston JR (1991) Rejoinder to discussion of Pilkey and Leonard (1990). J Coastal Res 7:565–577
Houston JR (1995) Beach replenishment. Shore Beach 63:21–24
Hydraulics Research Station (1970) Colliery waste dumping on the Durham Coast. Report 521. Wallingford, UK
Ibe AC, Awosika LF, Ibe CE, Inegbedion LE (1991) Monitoring of the 1985–86 beach nourishment project at Bar Beach, Victoria Island, Lagos, Nigeria. Proc Coast Zone 91:534–552
James WR (1974) Beach fill stability and borrow material structure. Proc Coastal 1334–1344
James WR (1975) Techniques in evaluating stability of borrow material for beach nourishment. US Army Coastal Engineering Research Center, Technical Manual 60
Jansen WA (1985) Shore stabilization and beach nourishment, Illinois Beach State Park. Shore Beach 53:3–6
Jelgersma S (1975) Netherlands. In: Bird ECF, Schwartz ML (eds) The world's coastline. Van Nostrand Reinhold, New York, pp 343–352
Johnson JW (1959) The littoral drift problem at shoreline harbours. J Waterways Harbors Div Proc Amer Soc Civil Eng 124:525–555
Jones JCE, Schafter IF (1986) Case studies of beach nourishment in Port Phillip Bay, Victoria, Australia. Manual on Artificial Beach Nourishment, Delft Hydraulics Laboratory, Netherlands, pp 26–36
Kana TW (1989) Erosion and beach restoration at Seabrook Island, South Carolina. Shore Beach 57:3–18
Kana TW, Stevens FD (1990) Beach and dune restoration following Hugo. Shore Beach 58:57–63
Kerckaert P, Paul PL, Roovers AN, De Cand TP (1986) Artificial beach nourishment on Belgian east coast. J Waterways Port Coast Ocean Eng 112:560–571
Kirk RM, Weaver RJ (1985) Coastal erosion and beach re-nourishment at Washdyke Lagoon, South Canterbury, New Zealand. Proc Australasian Conf Coast Ocean Eng 1:519–524
Komar PD (2007) The design of stable and aesthetic beach fills: learning from nature. Coast Sediments 7(2007):420–433
Krumbein WC (1957) A method for specification of sand for beach fills. Technical Memorandum 102, Beach Erosion Board, Corps of Engineers
Kuang CP, Pan Y, Zhang Y, Liu SG, Yang YX, Zhang JB, Dong P (2011) Performance evaluation of a beach nourishment project at West Beach in Beidaihe, China. J Coastal Res 27:769–783
Kumada T, Uda T, Serizawa M, Ishii H, Nakamura T, Oki Y (2007) Numerical simulation and field experiment on beach nourishment using coarse materials. In: Proceedings of 4th international conference on Asian and Pacific Coasts, pp 1094–1105
Kunz H (1990) Artificial beach nourishment on Nordeney, a case study. In: Proceedings of 22nd coastal engineering conference, pp 3254–3267
Leonard LA, Clayton T, Pilkey OH (1990a) An analysis of replenished beach design parameters on U.S. East Coast barrier islands. J Coastal Res 6:15–36
Leonard LA, Dixon KL, Pilkey OH (1990b) A comparison of beach replenishment on the U.S. Atlantic, Pacific and Gulf coasts. J Coastal Res 6:127–140 (special issue)
Louisse CJ, Kuik AJ (1990) Coastal defence alternatives in the Netherlands. In: Proceedings of 22nd conference on coastal engineering, pp 1.1–1.14
Macintosh KJ, Anglin CD (1988) Artificial beach units on Lake Michigan. In: Proceedings of 21st conference on coastal engineering, pp 2840–2854
Makowski C, Rusenko K (2007) Recycled glass cullet as an alternative beach fill material: results of biological and chemical analyses. J Coastal Res 23:545–552
Makowski C, Finkl CW, Rusenko K (2011) Using recycled glass cullet for coastal protection: a review of geotechnical, biological, & abiotic analyses. J Coastal Res 64:1362–1366 (special issue)
Marsh JK, Pillsworth MW, Kenny AJ (1997) Application of innovative fluorescent artificial sand and silt tracers for maintenance dredging optimisation, beneficial use and environmental dredging. In: Proceedings of 2nd Asian and Australasian ports and harbours conference 1997, Ho Chi Minh City, Vietnam, pp 173–188

McComb P, Black KP (2004) Detailed observations of littoral transport using artificial sediment tracer in a high-energy, rocky reef and iron sand environment. J Coastal Res 21:358–373

McFarland S, Whitcombe L, Collins M (1994) Recent shingle beach renourishment schemes in the UK: some preliminary observations. Ocean Coast Manag 25:143–149

Mocke R, Edwards A, Barr S (2005) A review of sand bypassing techniques for Bandy Creek Boat Harbour, Esperance. In: Proceedings of the 2005 coasts and ports conference, Australia, pp 131–136

Møller JT (1990) Artificial beach nourishment on the Danish North Sea coast. J Coastal Res 6:1–10 (special issue)

Moritz H, Puckette T, Marsh J, Boudreau R, Siipola M, Ott M (2011) Utilising sediment tracer studies to evaluate transport pathways at the mouth of the Columbia River, USA. In: Proceedings of coastal sediments conference, Florida, pp 1565–1579

Nankervis L (2005) Beach nourishment with the submarine sandshifter. In: Proceedings of the 2005 coasts and ports conference, Australia, pp 341–344

Neilson AF (1994) Subaqueous beach fluctuations on the Australian South-Eastern seaboard. Australian Civil Eng Trans CE36:57–56

Newman DE (1976) Beach replenishment: sea defences and a review of the role of artificial beach replenishment. Proc Inst Civil Eng 60:445–460

Nicholls RJ (1990) Managing erosion problems on shingle beaches: British experience. In: Proceedings of 3rd European Workshop on Coastal Zones, pp 2.35–2.36

Nordstrom KF (2000) Beaches and dunes of developed coasts. Cambridge University Press, Cambridge

Nordstrom KF, Allen JR (1980) Geomorphologically compatible solutions to beach erosion. Zeitschrift für Geomorphologie, Supplementband 34:142–154

Nordstrom KF, Allen JR, Sherman, DJ, Psuty NP (1979) Management considerations for beach nourishment at Sandy Hook, New Jersey. Coastal Eng 2:215–236

Noyce T (1993) Beach and offshore sand modelling in South Australia. G.I.S Section, Department of Housing and Urban Development, Adelaide

NRC (National Research Council) (1995) Beach nourishment and protection, commission on engineering and technical systems. National Academy Press, Washington D.C.

Nunny RS (1978) A survey of the dispersal of colliery waste from Lynemouth Beach, Northumberland. Technical Report 43, M.A.F.F., Great Yarmouth

Ojeda E, Ruessink BG, Guillen J (2008) Morphodynamic response of a two-barred beach to shoreface nourishment. Coast Eng 55:1185–1196

Oliveira JF, Muehe D (2013) Identification of the occurrence of compatible sediments for beach nourishment in the inner continental shelf between the towns of Niterói and Macaé—Rio de Janeiro, Brazil. J Integr Coast Zone Manage 13(1):89–99 (in Portuguese, English abstract)

Oliveira FSBF, Freire P, Sancho F, Vicente CM, Clímaco C (2011) Rehabilitation and protection of Colwyn Bay beach: a case study. J Coastal Res 64:1272–1276 (special issue)

Papps D, Priestley S (2005) Design and performance of replenishment projects on Auckland's eastern beaches. In: Proceedings of the 2005 coasts and ports conference, Australia, pp 125–130

Parry GD, Collett LC (1985) The control of seagrass in the Rosebud-Rye region of Port Phillip Bay. Coastal Unit Technical Report 2, Ministry for Planning and Environment, Victoria, Australia

Paskoff R, Petiot R (1990) Coastal progradation as a by-product of human activity: an example from Chañaral Bay, Atacama Desert, Chile. J Coastal Res 6:91–102 (special issue)

Pearson DR, Riggs SR (1981) Relationship of surface sediments on the lower forebeach and nearshore shelf to beach nourishment at Wrightsville Beach, N.C. Shore Beach 49:26–31

Pezzuto PR, Resgalla JRC, Abreu JGN, Menezes JT (2006) Environmental impacts of nourishment of Balneario Camboriu Beach, SC, Brazil. J Coastal Res 39:863–868 (special issue)

Pilkey OH (1990) A time to look back at beach replenishment. J Coastal Res 6:iii–vii

Pilkey OH (1995) The fox guarding the hen house. J Coastal Res 11:iii–v

Pilkey OH, Clayton TD (1987) Beach replenishment: the national solution? In: Proceedings of coastal zone '87 conference, New York, pp 1408–1419

Pilkey OH, Clayton TD (1989) Summary of beach replenishment experience on U.S. East Coast barrier islands. J Coastal Res 5:147–159
Pilkey OH, Leonard LA (1990) Reply to Houston discussion. J Coastal Res 6:1047–1057
Pluijm M (1990) Seaward coastal defence for the Dutch coast. In: Proceedings of 22nd conference on coastal engineering, pp 2010–2017
Poff M, Bass A, Sweeney R, Bahlinger K, Chatellier M (2011) Feasibility study analysis for riverine sand mining/Scofield Island restoration. J Coastal Res 59:235–245 (special issue)
Posford D (1998) Review of Lincshore renourishment strategy study. Report prepared for the Environment Agency, Anglian Region
Psuty NP (1984) USA New Jersey and New York. In: Walker H J (ed) Artificial structures and shorelines. Louisiana State University, pp 71–78
Psuty NP, Moreira ME (1990) Nourishment of a cliffed coastline, Praia da Rocha, the Algarve, Portugal. J Coastal Res 6:21–32 (special issue)
Ramana Murthy MV, Mani JS, Subramanian BR (2008) Evolution and performance of beach fill at Ennore Seaport, southeast coast of India. J Coastal Res 24:232–243
Riddell KJ, Young SW (1992) The management and creation of beaches for coastal defence. J Int Water Res Manage 139:588–597
Roberts TM, Wang P (2012) Four year performance and associated controlling factors of several beach nourishment projects along three adjacent barrier islands, west central Florida, USA. Coast Eng 70:21–39
Roellig DA (1989) Shoreline response to beach nourishment. Proc Coastal Zone 89:2104–2109
Roelvink JA (1989) Feasibility of offshore nourishment of Dutch sandy coast. In: International Conference on Hydraulic and Environmental Modelling of Coastal, Estuarine and River Waters, Bradford, pp 107–114
Royal HaskoningDHV (2009) Sand engine, Holland: environmental impact assessment
Schwartz RK, Musialawski FR (1977) Nearshore disposal: onshore sediment transport. Proc Coast Sediments 77:85–101
Schwartz ML, Marti JLJ, Herrera JF, Montero GG (1991) Artificial nourishment at Varadero Beach. Cuba Coast Sediments 91:2081–2088
She K, Trim L, Horn D A, Canning P (2007) Effects of permeability on the performance of mixed sand-gravel beaches. In: Proceedings of the 6th International symposium on coastal engineering and science of coastal sediment process, New Orleans, Louisiana, pp 520–530
Sherman DJ, Bauer BO, Nordstrom KF, Allen JR (1990) A tracer study of sediment transport in the vicinity of a groin: New York, USA. J Coastal Res 6:427–438
Shuisky YD (1994) An experience of studying artificial ground terraces as a means of coastal protection. Ocean Coast Manag 22:127–139
Silvester R (1974) Coastal engineering. Elsevier, Amsterdam
Silvester R (1976) Headland defense of coasts. Coast Eng 1:1394–1406
So CL (1974) Some coast changes around Aberystwyth and Tanybwlch, Wales. Trans Inst Brit Geogr 62:143–153
Stive MJF, Nicholls RL, De Vriend HJ (1991) Sea level rise and shore nourishment: a discussion. J Coast Eng 16:147–163
Stive M, Schipper M, Luijendijk A, Theil de Vries J, Aarninkhof S, Gelder-Maas C, Theil de Vries S, Henriquez M, Marx S, Ranasinghe R (2013) A new alternative to saving our beaches from sea-level rise: the sand engine. J Coastal Res 29:1001–1008
Sudar RA, Pope J, Hillyer T, Crumm J (1995) Shore protection projects of the U.S Army Corps of Engineers. Shore Beach 63:3–16
Thompson CL (1987) Beach nourishment monitoring on the Great Lakes. Proc Coast Sediments 87:1203–1215
Uda T, Naito K, Kanda Y (1991) Field experiment on sand bypass of the Iioka Coast. Coast Eng Jpn 34:205–220
U.S. Army Corps of Engineers (1994) Shoreline protection and beach erosion control study—Phase 1: cost comparison of shoreline protection projects of the U.S. Army Corps of Engineers
US Army Corps of Engineers (USACE) (1991) Sand bypassing system, engineering and design manual. Washington DC

US Army Corps of Engineers (USACE) (2002) Coastal engineering manual. Washington DC
Usoro EJ (2010) Nigeria. In: Bird ECF (ed) Encyclopedia of the world's coastal landforms. Springer, Dordrecht, pp 949–952
Vera Cruz D (1972) Artificial nourishment at Copacabana Beach. In: Proceedings of 13th coastal engineering conference, pp 1451–1463
Verhagen HJ (1992) Method for artificial beach nourishment. In: Proceedings 23rd International Conference on Coastal Engineering, Venice, Italy, pp 2474–2485
Walton TL, Dean RG (1976) Use of outer bars of inlets as sources of beach nourishment material. Shore Beach 44:13–19
Walton TL, Purpura JS (1977) Beach nourishment along the southeast Atlantic and Gulf coasts. Shore Beach 45:10–18
Weggel JR (1995) A primer on monitoring beach nourishment projects. Shore Beach 63(3):20–24
Weggel JR, Sorensen RM (1991) Performance of the 1986 Atlantic City, New Jersey, beach nourishment project. Shore Beach 55:2–11
Williams AT, Micallef A (2009) Beach management: principles and practice. Earthscan, London
Willis DH, Price WA (1975) Trends in the application of research to solve coastal engineering problems. In: Hails JR, Carr AP (eds) Nearshore sediment dynamics and sedimentation. Wiley, New York, pp 111–122
Willmington RH (1983) The renourishment of Bournemouth beach 1974-1975. In: Proceedings of shoreline protection conference, September 1982, Telford, London, pp 157–162
Wynne AA (1984) Adelaide coast protection strategy review. Coast Protection Board, South Australia
Zenkovich VP, Schwartz ML (1987) Protecting the Black Sea-Georgian SSR gravel coast. J Coastal Res 3:201–209
Zunica M (1990) Beach behaviour and defences along the Lido di Jesolo, Gulf of Venice, Italy. J Coastal Res 6:709–719

Chapter 5
Beach Renourishment for Coast Protection

Abstract Coastal protection is one of many objectives of beach renourishment and is considered in this chapter as a background to the debate between the traditional use of structures and alternative 'soft' or adaptive options for coastal management (and protection). A number of examples are used to illustrate some of the principles when considering beach renourishment for coastal protection.

The traditional response to coastal erosion has been to build solid structures such as sea walls or boulder ramparts, to protect the coastline, but it has been realised that a renourished beach that prevents storm waves from attacking the base of a cliff can be as effective a means of coastal protection as solid structures, providing it persists for a sufficient period to be cost-effective.

Renourished beaches have sometimes been added in front of previously built sea walls, or inserted between groynes, as a supplementary means of coastline protection, to make the coastline less artificial in appearance and to provide a recreational resource. Addition of a renourished beach on the seaward side of a sea wall has been seen as a way of 'softening' hard engineering at several sites on the Netherlands coast, and at Melaka in Malaysia.

In recent years there have been several projects using renourished beaches as an alternative to engineering works to reduce erosion and coastal flooding. In the United Kingdom beach renourishment on the Lincolnshire coast has been introduced since the early 1990s to protect land and property at risk from flooding (Sect. 4.2.7, p. 49). The beaches and dunes stabilised by renourishment currently protect 35,000 ha of urban and agricultural land and over 16,000 residential, 1,700 commercial/industrial properties, 19,000 static caravans and various environmental assets, as well as coastal resorts. The renourishment is estimated to protect the hinterland against inundation from up to a 1 in 200 year tidal flood.

At Sand Bay in Somerset beach renourishment in 1984 proved successful in stabilising the beach and dunes to protect large areas from extensive flooding until

E. Bird and N. Lewis, *Beach Renourishment*, SpringerBriefs in Earth Sciences,
DOI 10.1007/978-3-319-09728-2_5

the unusual floods of 2014. Further protective works, including more beach renourishment, are now being considered.

On the north-east coast of Port Phillip Bay, Australia, a sector of natural vegetated bluffs south from Quiet Corner had been stable until the beach fronting them was depleted following the completion of a masonry sea wall at Black Rock, to the north, in 1939. Reduction of the beach allowed waves from the west to generate stronger longshore drift, so that the beach fronting the bluffs gradually dwindled. By the 1960s storm waves were undercutting these bluffs, and erosion was threatening to undermine this part of the coastal highway (Fig. 5.1).

A proposal to extend the sea wall to halt this cliffing was opposed by local residents, who argued that the beach should be renourished to prevent storm waves reaching the base of the bluff, which would also restore scenic and recreational values. In response, a beach terrace 100 m long, 25 m wide, and a metre above high spring tide level, was formed by pumping coarse shelly sand in from the sea floor during the winter of 1984 (Fig. 5.2). This coast shows a seasonal alternation in longshore drift, and because of the SSW aspect of the shore the longshore drift is stronger to the SE in the summer half-year than to the NW in the winter half-year. The outcome was that sand lost from the renourished beach south of Quiet Corner was carried south-eastward each winter by longshore drift by waves arriving from the west, so that as this beach became narrower, while the beach SE to Banksia Point and beyond widened (Fig. 5.2a, b). There was little north-westward drift in the summer, so the emplaced beach did not grow in that direction (X on Fig. 5.2c). Successive profile surveys showed that the emplaced beach was also cut back by storm waves, some of the finer sand withdrawn from the beach being deposited as a sand bar that persisted in the nearshore zone, the crest of which moved shoreward in calmer weather and seaward during storms (Fig. 5.3). Although somewhat depleted, this renourished beach has remained in position for 30 years, and has served its purpose of protecting the bluff base from storm wave erosion as well as maintaining a recreational resource.

The sandy beach between Picnic Point and Red Bluff at Sandringham, on the NE coast of Port Phillip Bay, stands in front of vegetated bluffs. As on the other

Fig. 5.1 The shore south of Black Rock in 1968, when storm waves at high tide were eroding the base of the backing bluff. © Geostudies

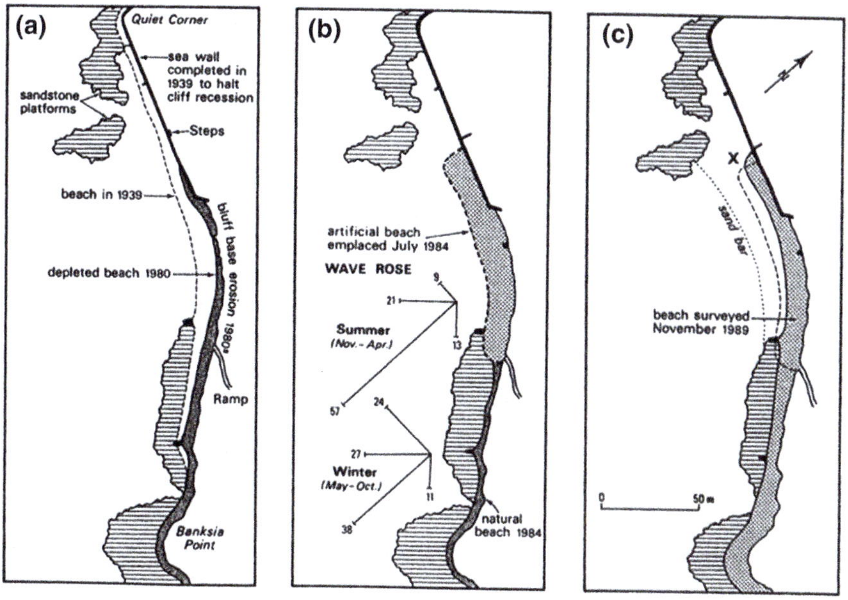

Fig. 5.2 Changes on the coast south of Quiet Corner, Black Rock, Port Phillip Bay. **a** Depletion of beach after sea wall construction, **b** renourishment of beach in 1984, **c** beach sand extended southward past Banksia Point, but showed little change at the northern end (X). © Geostudies

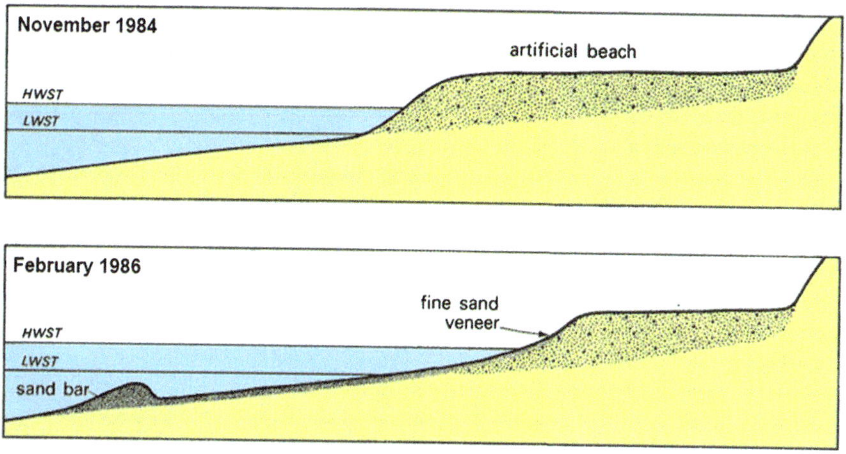

Fig. 5.3 The change in profile on the renourished beach south of Quiet Corner, Black Rock, Port Phillip Bay. © Geostudies

beaches on this coast northward drift of sand widens the beach near Picnic Point in the summer half-year and there is southward drift towards Red Bluff during the winter. Air photographs taken in 1930 and subsequently show that the beach has been gradually depleted in recent decades, and by 1989 the southern part had become very narrow in winter, when the backing bluff was undercut by storm wave erosion, forming slumping cliffs of sandy clay up to 10 m high (Fig. 5.4). As these were cut back, there was a risk that a segment of the coastal highway, which here runs close to the top of the bluffs, would be undermined.

In 1990 a groyne was built at Edward Street and the sector south to Red Bluff renourished to form a beach 25 m wide and 600 m long by trucking in about 35,000 m^3 of sand, placed to protect the base of the bluffs from further storm wave erosion (Fig. 5.5). As a result the undercutting of the bluffs ceased (Fig. 5.5). Some slumping continues because of groundwater seepage from the bluff, but by 2004 the cliff base had become largely revegetated. Protection of a cliff base by

Fig. 5.4 Erosion of backshore bluff on a sector of Sandringham Beach, Port Phillip Bay in 1989. © Geostudies

Fig. 5.5 After the renourishment of the of Sandringham Beach, Port Phillip Bay, south of the Edward Street groyne, in 1990, backshore cliff erosion halted and vegetation revived. © Geostudies

means of a renourished beach will be effective as long as that beach is maintained, if necessary by periodic renourishment.

The beach was thus restored in the southern third of the Sandringham beach compartment, but north of the Edward Street groyne it remained narrow, and occasional storm waves began to undercut the backing bluff. In 2004 there was discussion on whether this could be controlled by emplacing another protective artificial beach, or whether a solid sea wall would be built. Critics of the sea wall proposal pointed out that the fronting beach would be further depleted by reflection scour. Seasonal alternations of longshore drift continued, and in the winter months sand drifted southward to form a protective beach in front of the eroded cliffs, but in the summer it drifted away, exposing the cliffs to renewed wave attack.

In 2008 a groyne was built opposite Southey Street at the northern limit of cliff erosion, and sand from Sandringham Harbour was piped a kilometre southward to renourish the beach between the new groyne and the one previously built at Edward Street (Fig. 5.6). This has stabilised the coast in the central part of the Sandringham beach compartment, but the northern part now shows beach depletion and incipient cliffing of the backing bluff, and it may yet be necessary to add a further groyne and renourish this next section.

The insertion of groynes in the Sandringham beach compartment led to protests by some local people (despite the fact that groynes had been introduced successfully

Fig. 5.6 The renourished beach between groynes at Sandringham, Port Phillip Bay, in 2010. © Geostudies

in the Hampton beach compartment to the north: Sect. 4.3.2, p. 65). Some thought that it would have been better to renourish the whole of the Sandringham beach compartment without inserting any groynes. It would indeed be possible to do this, given a sufficiently large quantity of sand, but the effects of seasonal longshore drift must be taken into account. Each winter sand drifts northward and would accumulate alongside Picnic Point, spilling round into Sandringham Harbour (in the same way that the renourished beach between New Street and Green Point lost sand northward round Green Point in successive winters: Sect. 4.3.10, p. 76). Sandringham Harbour already has a problem of excessive sand accretion, and each summer southward drift of sand would widen the beach towards Red Bluff until sand spilled round into the next bay. These losses from the extremities of the Sandringham beach compartment would in due course deplete the renourished beach until storm waves resumed their attack on the base of the backing bluff.

Seasonal alternations of longshore drift thus make it difficult to maintain a renourished beach in an elongated beach compartment, and it is necessary to introduce groynes to divide the compartment into manageable sections.

These projects have demonstrated the importance of renourishing and maintaining a wide, high and persistent beach to prevent cliff recession. Such a beach is a means of absorbing wave energy and protecting the coastline from further erosion. It is important that a sufficient volume of beach material be maintained to protect the backshore, because a small quantity of sand or shingle that can be mobilised by storm waves can actually accelerate abrasion of cliffs or solid structures. This was the cause of severe erosion when the shingle beach was depleted at Hallsands in south-west England. Increased abrasion occurred on the sea wall at Aberystwyth in Wales after much of the sediment used in a beach renourishment project quickly weathered and dispersed, leaving small quantities of hard granitic gravel that were hurled at the wall by storm waves (So 1974).

Reference

So CL (1974) Some coast changes around Aberystwyth and Tanybwlch, Wales. Trans Inst Brit Geogr 62:143–153

Chapter 6
Planning Considerations

Abstract A number of key considerations that need to be taken into account when planning beach renourishment are introduced. Environmental impacts, both positive and negative are discussed, along with the costs of beach renourishment. Predictions of global warming and a world-wide sea level rise are considered and the implications for beach renourishment projects discussed.

6.1 Environmental Impacts of Beach Renourishment

Beach renourishment is generally beneficial, and widely regarded as a better alternative to the construction of hard structures (Adriaanse and Coosen 1991; Hamm et al. 2002; Finkl 2002), but it can produce adverse impacts on the environment (Speybroeck et al. 2006). When works are undertaken the activity is often intensive (Fig. 6.1).

Beach renourishment requires the extraction of suitable fill from a source area, its transportation to the shore and its deposition on a beach. Each of these procedures may have negative environmental impacts, both during construction and once the construction activity has ceased. Impacts can include sediment disturbance leading to increased turbidity in nearshore waters and waters close to the sediment source, the displacement or burial of plant and animal communities, and associated changes in oxygen, temperature, salinity, light penetration, and the circulation of nutrients and chemicals in the sea and on the sea floor. Marine plant and animal communities may be reduced or destroyed, and their revival may be very slow, depending on the degree of disruption and the availability and vigor of recolonising biota (Pullen and Naqui 1983).

Negative effects of beach nourishment dominate in the short to medium term, the size of the impact being determined by (1) activities during the construction phase, (2) the quality and the quantity of the nourishment sand, (3) the timing, place and size of project, and (4) the nourishment technique applied (Speybroeck

E. Bird and N. Lewis, *Beach Renourishment*, SpringerBriefs in Earth Sciences,
DOI 10.1007/978-3-319-09728-2_6

Fig. 6.1 Beach renourishment on the Lincolnshire coast, United Kingdom. Between 2010 and 2015 15,000 m^3 of sediment per day was pumped ashore and placed on the beach using hydraulic pipes. Sediment was then spread across the beach using excavators and bulldozers. © Dredging International

et al. 2006). Some of these effects can be reduced by planning the timing of dredging, transportation and delivery of sediment to renourish beaches. This depends partly on favorable weather conditions, and should also avoid breeding seasons so that the impact on biota is less severe. The preferred time for renourishment depends on the nature, location and the species inhabiting the site.

The assessment of potential environmental impacts is undertaken for large renourishment projects, but the statutory requirement changes from country to country and is usually determined by the construction time frame or project size (placement quantity). For example, in Spain an Environmental Impact Assessment is a legal requirement for projects involving a placement of more than 500,000 m^3, while in the United Kingdom environmental assessments are undertaken to a level of detail proportionate to the project size. In some instances an assessment of the potential environmental impacts is completely lacking. For example, renourishment of the beach at Balneario Camboriu in 2002 was not preceded by any specific environmental assessment, yet 50,000 m^3 of material was hydraulically dredged and placed along 800 m of beach. The renourishment caused a number of impacts including significant changes to the beach and bay sedimentology, death of macrofauna due to the dredging operations and suffocation from placement, and suffocation of filter feeders such as bivalves (Pezzuto et al. 2006).

A comprehensive assessment of environmental impacts would consider a number of environmental parameters, including ecology, water and sediment quality, geology and geomorphology, fisheries, socio-economics, landscape, archaeology, navigation, air quality and noise.

The impacts of beach renourishment have often proven difficult to measure, particularly indirect impacts that happen away from the project area and cumulative impacts of multiple renourishments (Peterson and Bishop 2005). A better understanding of the potential environmental impacts of beach renourishment requires an accurate description of prior environmental conditions. As long-term data on natural fluctuations in populations of marine organisms, due to storm waves, winter mortality for example, are often not available, an assessment of the effects of beach renourishment projects may be difficult to complete (Herrera et al. 2010).

6.1.1 Source Impacts

Dredging of sediment from the sea floor has been widely used as a source of sand and gravel for building, road-making and other constructional work as well as sediment for beach renourishment. Such dredging disrupts sea floor ecosystems, in particular submarine vegetation such as seagrass beds. It is necessary to select areas for sea floor dredging that are well away from critical habitats, breeding and feeding areas. Biological surveys should be made before dredging begins, and sensitive areas such as coral reefs, seagrass areas, and habitats for fish and shellfish mapped in order that they can be avoided.

Increased sediment in the water column and increased sedimentation is known to negatively influence corals (Goldberg and Wilkinson 2004). Dredging associated with renourishments in Broward County, Florida showed a localised effect on the sediment regime, with dredging near coral reefs causing higher rates of sedimentation than elsewhere (Jordan et al. 2010).

Extraction of nearshore sand to renourish a recreational beach in front of the Promenade de la Plage at Prado, near Marseille in southern France between 1974 and 1982 led to destruction of nearshore *Posidonia* beds (Rouch and Bellessort 1990). Eroding beaches on the shores of Hel spit on the coast of Poland were renourished with sand dredged from the floor of Puck Bay, a lagoon to the south, and pumped across the spit, but this was stopped when it became clear that the dredging was damaging vegetation, increasing turbidity, and reducing fish populations in the lagoon. Sand was then obtained from deposits on the floor of the Baltic Sea to the north and pumped into the shore (Basinski 1994).

It is necessary to ensure that sea floor excavations do not excessively deepen nearshore areas, because this can lead to increases in wave energy, initiating or accelerating coastal erosion. Benedet et al. (2013) assessed the effects of nearshore dredge pits on adjacent beaches, showing that dredge pits can influence waves, currents, and sediment transport, but demonstrated that impacts can be reduced by adapting pit designs without having to reduce the overall dredging volume.

Deep excavations can become stagnant hollows, which are anaerobic and ecologically unproductive. Shallow dredging over a larger area may be initially more damaging, but ecological recovery is much quicker on shallow excavations. This was illustrated during beach renourishment on the island of Sylt (Sect. 4.3.2, p. 62), when sand was excavated from the nearshore sea floor by a hopper dredge, which cut to a depth of about 2 m and disturbed large areas. Deeper dredging from an anchored hopper could restrict disturbance to a much smaller area by cutting to a depth of 40 m, but it was decided that deep excavations could have more severe adverse impacts on wave processes and marine ecosystems (Dette 1990).

When sea floor sources of sand were sought for the renourishment of Mentone Beach, Port Phillip Bay, Australia, in 1976 (Sect. 4.2.7, p. 49) there were fears that the dredging of sand would have adverse effects on sea floor ecosystems, notably seagrass communities, a habitat for fish and shellfish. However there was reassurance from Watson (1973), who had found that excavation of a trench 5 m wide and up to 3 m deep to carry a gas pipeline across the floor of Port Phillip Bay in 1972 had caused only temporary depletion of benthic organisms, and may even have enriched the local fishery. In the event, the replenishment of Mentone beach does not appear to have adversely affected the sea floor ecosystems in Port Phillip Bay, although corrosion of the pipeline may yet pose a problem.

Dredging of sediment from the sea floor may release toxic chemicals. At Bogue Banks, North Carolina, sediment dredged from a harbour was found to be laden with hydrogen sulphide, which caused much intertidal and nearshore turbidity, modifying the habitat and killing many invertebrates. The sea floor ecosystem began to recover only slowly after beach dumping ended (Reilly and Bellis 1983).

Monitoring of sea floor plant and animal communities has shown that many gradually recover after dredging has ceased. In Florida surveys showed good recovery of sea floor biota 5 years after the dredging areas off Hillsboro Beach (Marsh and Turbeville 1981) and in the Tyrrhenian Sea, Italy, analysis of the effects of dredging on macrobenthic fauna demonstrated that although the richness and diversity of species was reduced during dredging re-colonisation could be observed a few months after dredging ceased (La Porta et al. 2009). This study also found that re-colonisation was more rapid for areas after one dredging, compared with areas after two, agreeing with other studies (Cooper et al. 2007) that the re-colonisation of benthic assemblages is related to the intensity of dredging.

Sea floor ecosystems may also be damaged by burial or increased turbidity when sediment dredged from harbours or harbour approaches is dumped offshore. It may be better to use dredged sediment for beach renourishment or land reclamation instead of dumping it offshore, where it can have ecologically adverse impacts on the vegetation that sustains the sea floor fauna, including fisheries.

6.1.2 Impacts During Transportation

Sediment dredged from the sea floor has to be transported to the shore, either in pipes or on boats. Leakages from pipes during pumping or losses as boats are loaded, navigated and unloaded can cause turbidity in the sea, the coarser gravel and sand settling quickly but the fine-grained silt and clay remaining in suspension. Cloudy water can diminish light penetration and so disadvantage sea floor vegetation, while blanketing by spilt sediment has damaged seagrass communities, coral reefs, and fish and shellfish resources.

Some of the early beach renourishment projects in the United States used sediment obtained from dredging nearby lagoons. These contained high proportions of fine-grained sediment, the release of which buried, or proved damaging to, sea floor biota (Reilly and Bellis 1983). Later use of coarser sediment from offshore has caused less damage to estuarine and nearshore ecosystems (Marsh and Turbeville 1981; Lankford and Baca 1989).

Mention has been made of problems of overland transportation, particularly lorry traffic passing through seaside resorts.

6.1.3 Impacts of Beach Emplacement

The sand beach is a productive habitat, supporting dense concentrations of benthic invertebrates that feed fishes, shore birds and crabs (Brown and McLachlan 1990). Sea turtles nest on some beaches. Renourishment of beaches can have ecological impacts, as when previously rocky or muddy shore habitats are buried beneath sand. The impacts of beach renourishment vary with location: offshore nourishment will mainly affect benthic species and foraging birds, whereas backshore renourishment will impact on terrestrial plants and animals (Speybroeck et al. 2006).

Beaches, particularly sandy beaches, are ecosystems adapted to natural changes caused by cut and fill and longshore drift, as well as tidal oscillations and frequent variations in wave energy and turbidity. Ecosystems are modified as habitat diminishes on eroding beaches, or as renourishment proceeds. Studies of ecological changes on renourished beaches require monitoring of organisms on beaches before and after renourishment (Nelson 1993; Adriaanse and Coosen 1991). At Myrtle Beach in South Carolina sand quarried from inland was trucked to the shore in early spring, and its deposition caused initial reductions in beach organisms, but there was then rapid recovery and after four months some sites actually showed species enrichment (Baca and Lankford 1988).

At Palm Beach on Australia's Gold Coast macrofauna sampling was conducted before and after the placement of 30,000 m^3 of sediment in November 2007. It was found that macrofaunal communities were disrupted by the placement of sand and associated works on the area (Noriega 2008). Five months after renourishment macrofauna improved and returned to pre-renourishment levels. Surf zone fish and shore birds feed on macrofauna, so a decline in macrobenthic communities during beach renourishment could have an adverse effect on these species.

The impact of beach renourishment on turtles in the United States has been discussed by Dean (2002). Adverse impacts can include a harder and more compacted beach due to the presence of finer sediment and a different sand colour, which affects the incubation temperature. The renourished beach can include higher berms which can make it difficult for turtles to move up the beach to favourable nesting sites.

Increased accretion downdrift from renourished beaches may have adverse effects on ecosystems, as at Saintes Maries de la Mer, on the southern coast of France, where sand is drifting on to the Camargue salt marshes. At Rapid Bay in South Australia sand and gravel eroded from a beach formed by quarry waste dumping has drifted northward across the sea floor, blanketing formerly rich ecosystems on reefs and impoverishing the local fishery (Bourman 1990). Sediment can also be eroded and moved offshore, impacting on nearshore habitat such as seagrass. Reference has been made to the proliferation of nearshore seagrass as a consequence of bulldozing sand up on to the beach at Rosebud, Victoria, Australia (Sect. 4.3.6, p. 71).

Beach renourishment can increase the suspended sediment concentrations of adjacent waters, both at the time of sand placement and subsequently as sediments are redistributed. During the renourishment of Swanage, Poole and Bournemouth beaches in 2006, of a total volume of sediment pumped ashore 30 % was lost from the beach (CIRIA 2010). Several factors can contribute to the amount of sediment suspension during and after placement on the beach, including the mode of placement, meterological conditions and the proportion of finer sediment contained in the renourished beach. Higher concentrations of suspended sediment cause increased turbidity, but this a temporary effect and is often dispersed quickly by wave action (Van Dolah et al. 1992). Furthermore, nearshore biological communities have a natural resilience to shifts in turbidity, which is a natural phenomenon (Van Dolah et al. 1994). Wilber et al. (2006) recorded higher suspended sediment concentrations in the swash zone adjacent to the renourishment site compared with other areas, but little difference in the surf zone or nearshore. The study also found higher suspended sediment extensive following storms than after beach renourishment. Studies conducted by Rakocinski et al. (1996) during and after extensive beach restoration at Perdido Key, Florida, demonstrated important changes in benthic structure in response to silt/clay loading in the nearshore and offshore areas, more than 2 years after the end of the project. Responses of the fauna included decreased species richness and total abundance.

Substantial amounts of sediment dredged from the sea floor for use in beach renourishment can be lost during delivery to the shore. The Rockaway Beach project in 1975–1977 lost 10 % of the volume of sediment originally excavated from the sea floor before it reached the beach. Some was lost during dredging, some as the barges were filled, and some as they were emptied on to the shore. Such losses, mainly of fine-grained sediment, modify the grain size distribution of the material that reached the beach. At Rockaway Beach the losses of fine-grained sediment were not a problem, as relatively coarse sediment was required for beach renourishment there.

On the island of Sylt (Sect. 4.3.2, p. 62) a million m^3 of sand was extracted from the sea floor, but more than 20 % of it (also mainly the fine fraction) had been lost before the balance of 770,000 m^3 actually reached the shore (Dette 1977).

At New River, North Carolina a loss of 16 % of the original renourishment left coarser and better-sorted sediment, which helped to improve the performance of the restored beach (Hobson 1977).

The short-term construction phase can render the beach unusable, which in turn can have economic implications. Changes to the beach following renourishment may have impacts on recreational use. Where beaches are renourished with coarser sediment than the native sediment, usually in an attempt to improve durability, the resulting beach profile can be steeper. This coupled with the coarser sediment size make the beach less attractive to beach users.

Renourishment may also cause changes to nearshore bathymetry and wave breaking patterns, with implications for surf-riding conditions. This has been considered on a number of Australia's east coast beaches, for example Collaroy-Narrabeen, north of Sydney, where decisions on the methods of coastal management were influenced by their potential impacts on surf conditions, despite previous renourishments making use of sediment dredged from the adjacent lagoon entrance. Pitt (2012) compared a number of case studies from around the world where beach renourishment projects had resulted in both negative and positive impacts for surf riding conditions. At Scheveningen in the Netherlands, beach renourishment steepened the neashore profile and covered sandbars, resulting in poorer surfable conditions. In 2010 70,000 m^3 of sand was used to shape a point extending 100 m offshore. The sand bar which created plunging waves provided suitable surfing for up to 3 months. At Long Branch New Jersey, United States, renourishment was designed to improve conditions for surfing, but failed. At Aramoana Beach, New Zealand, offshore disposal of dredged sediment unexpectedly improved surfing opportunities. At Cronulla, Australia, beach nourishment included the offshore placement of sediment with the aim of improving surf conditions, and this was successful, resulting in more pronounced wave refraction, wave amplification and wave focusing.

Artificial coastlines have become extensive in Singapore, where land reclamation increased the area of the island by nearly 25 % between 1960 and 2010. On the northern (Strait of Johore) coast at Jurong and Changi some sectors were reclaimed by dumping earth and weathered rock, leaving their seaward edges unprotected, so that wave action could sort the sediment and form beaches. Longshore drift carried some of the reworked sediment to downcoast sectors, where beaches were improved if the sediment received was sandy, but spoiled where they were blanketed with silt and clay (Wong 1985).

6.2 Costs and Benefits of Beach Renourishment

Beach renourishment is costly, but may be economically justifiable on sectors of the coastline, such as seaside resorts, where the emplaced beaches will be much used, or where the beach will protect property or infrastructure that is at risk from erosion or flooding.

There are costs in seeking sources of beach material suitable for renourishment, in extracting, transporting and emplacing it on the shore, and in subsequent maintenance. Additionally, beach renourishment requires costs associated with design, licensing, approvals, as well as pre- and post-construction monitoring.

On the Atlantic coast of the United States the cost of beach renourishment and maintenance in 1995 was US$500,000 per (2014 US$700,000) mile per year. A similar figure was reported by Trembanis et al. (1999), who estimated the cost of maintaining nourished beaches for 10 years along the developed coastlines of New Jersey, North Carolina, South Carolina, and Florida would be $5.9 million per mile; New Jersey having the highest cost of $17.5 million per mile and South Carolina having the lowest, $3.3 million per mile (2014 US$7.6 million per mile, 2014 US$28 million per mile and 2014 US$4.2 million per mile respectively). They also found that the average cost per cubic yard was approximately US$5 (2014 US$6.5/cubic yard), but this also differed along the coastline. It was suggested by Campbell and Benedet (2006) that by 2006 the cost of beach renourishment in the United States was about US$12 per cubic metre (2014 US$13 per cubic metre). In the United Kingdom cost estimates for three beach renourishments in Wales ranged from £8 to £16 per cubic metre (Wellard and Rimington 2013).

It is also useful to compare costs for complete projects. Mention has been made of the ongoing beach renourishment on the Lincolnshire coast in the UK between Mablethorpe and Skegness (Sect. 4.2.7, p. 49), where placement of 500,000 m^3 per year between 2009 and 2013 has cost £6 million annually. At Ettalong Beach in Australia, renourishment was estimated to cost $AUS 500,000 per 30,000 m^3 renourishment.

The cost of beach renourishment can vary, depending on a number of variables. For example, an average beach renourishment in Florida is roughly half the cost of that in Massachusetts (Hoagland et al. 2012). Much depends on the available sources of material for beach renourishment and the distance across which suitable material must be conveyed. Renourishment of the beach at Bournemouth in 1974–1975 with sand dredged from a site several miles offshore cost over £1 million, but in 1988–1989 further renourishment with sand supplied as a product of dredging the adjacent tidal entrance to Poole Harbour cost only £130,000. The coincidence of a need for dredging with a nearby demand for beach renourishment thus resulted in substantial savings for Bournemouth.

Costs can be reduced by a number of means. Wellard and Rimington (2013) demonstrated that the unit cost of a beach renourishment could be reduced by 25 % by changing the volume of the dredge vessel. Economies of scale can be realised with greater volumes of material (Hoagland et al. 2012). Simplified placement methods can also reduce costs. Placing renourishment sediment to a selected height on the subaerial beach only, and allowing natural sorting rather than labour intensive re-profiling, can reduce the cost of renourishment (Clarke and Brookes 2008). This was demonstrated at Ettalong Beach, New South Wales (Sect. 4.3.4, p. 68) in 2013, where renourishment sediment was heaped on the beach and left to be sorted by wave action (Fig. 6.2). In Lincolnshire cost efficiencies were made by surveying

Fig. 6.2 Renourishment of Ettalong Beach, Broken Bay, Australia during November 2013. Sediment dredged from the bay floor 1 km offshore is pumped onshore through a hydraulic pipe and then placed in heaps along the sub-aerial beach by excavator. © Nick Lewis

the beach annually and comparing the result with the target profile before planning more specifically where, when and how best to undertake subsequent work.

Long-term costs of beach renourishment projects are difficult to estimate. The beach renourishment at Ocean City, New Jersey in 1982 (Sect. 4.3.8, p. 73) did not provide the expected benefits; it cost $2.5 million but lasted only two and a half months (NOAA 2000). As previously mentioned, reasons for such discrepancies in performance can include poor project design, unanticipated coastal storm events, or use of incompatible sand grain size. The unknown life of the rebuilt beach and the resulting need for maintenance renourishment are two causes of uncertainty in determining long-term costs and benefits.

Coastal planners should compare estimates of the costs of beach renourishment to the costs of alternatives, such as solid structures or managed retreat. The costs of beach renourishment are generally lower and more evenly spread over time than those incurred with the building of solid structures, which also require maintenance, especially after they are damaged by storms. Renourished beaches are more flexible than artificial structures because the beach profile can adapt to hydrodynamic variations, such as cut-and-fill sequences or storm events, without the damage caused to sea walls, groynes and other structures. In Port Phillip Bay, Australia, the cost of beach renourishment does not exceed that of building and maintaining solid structures if the renourished beach survives for more than 7 years, while, on the Georgian Black Sea coast beach renourishment for 51.9 million roubles (2014 US$1 million) was about half the cost of the previously unsuccessful coast protection works using solid structures. On the east coast of the United States the cost of maintaining a beach front lot through renourishment is around US$10,000 (2014 US$13,000) per year (Pilkey and Hume 2001), whereas sea wall construction costs about US$3,000 (2014 US$4,000) per metre, with maintenance costs of 4–10 % per annum, depending on exposure to wave action (Neumann and Livesay 2001). In reality, such comparisons are far more complex, and can only be effective when undertaken on a site- by-site basis.

The chief benefits of beach renourishment are the provision of improved scenic and recreational values and additional coastline protection against the effects of storms. There is a reduction in cliff erosion and storm damage on coastal structures such as esplanades, roads and buildings. Unlike sea walls and groynes, a renourished beach protects one sector without inducing erosion downdrift, and some of the sediment deposited may be carried by longshore drift to downdrift sectors, augmenting their beaches and thus improving protection for adjacent developed coasts and their communities. Sediment moving alongshore is not really 'lost' if it benefits adjacent beaches and coasts. Examples of this have been noted on the coast east of Bournemouth, and in Singapore, while on the south coast of Port Phillip Bay the erosion of the beach at Portsea (Sect. 2.7, p. 18, 19) has been accompanied by accretion on beaches downdrift towards Sorrento: by the end of 2013 Shelly Beach and Point King Beach at Sorrento were both exceptionally wide because sand lost from Portsea beach had drifted eastward along the shore (Fig. 2.10). Losses from gravel beaches emplaced on the shores of Lake Michigan are mainly alongshore rather than offshore, with the benefit that they may widen beaches that protect sectors downdrift along the coast (Roellig 1989).

Beach renourishment improves the recreational resource by increasing beach area. The widened beach is attractive to visitors because it is a more natural and pleasant environment for recreation than a coastline dominated by sea walls, tetrapods, breakwaters and groynes. A successful beach renourishment project provides a seaside resort with a more attractive tourist lure, and results in more visitors and increased income, compared with resorts that are losing their beaches or have become excessively adorned with artificial structures (Dean 1987).

Beach renourishment is easily justified in the Netherlands, where one-third of the land is below mean sea level and potentially massive socio-economic consequences are attributable to a rising sea level. It is a densely populated country (494 people/km^2) with a 350 km long coastline, and nine million residents (out of a total of 16.7 million) living in coastal areas, many of which are below mean sea level. Roughly 65 % of the country's gross national product (about €400 billion) is generated within this coastal region (Stive et al. 2013).

There is also the question of who should pay for beach renourishment projects. In most countries the cost is met by national or local government agencies on the grounds that the restored beaches are public facilities, but where the beaches are private, or public access is impeded for one reason or another, this becomes difficult to justify. In the 1970s beach erosion at Miami led to demands for public help, but by then 95 % of the beach had passed into private ownership. The United States Army Corps of Engineers decided to make unimpeded public access a condition of beach renourishment at public expense.

It has been found that renourishment of beaches on the east coast of the United States has been of benefit to property owners because it provides more effective protection from storm damage and gives them an improved recreational resource (Olsen 1982). On Tybee Island, Georgia, Landry and Hindsley (2011) showed that increased width of beaches and dunes augments property values within 300 m of the coastline. Prices of land and housing have risen, as do incomes from rent and tourist expenditure, behind beaches widened by renourishment in South Carolina (Pompe and Rinehart 1994) and North Carolina (Gopalakrishnan et al. 2011). On the Atlantic coast generally beach renourishment can increase real estate values by up to 21 % (Black et al. 1988), and prompt further development or rehabilitation of existing development (Stronge 1990; Bodge 1991).

Nevertheless, many coastal engineers remain cautious about the economic viability of beach renourishment projects. In the words of the British engineer, Barrett (1989): 'I have no doubt that there is a consensus amongst coastal engineers that the ideal form of coastal defence in purely engineering terms is a massive beach. Whether this can be achieved in future economic terms and in proper long-term use of available resources are quite another matter'.

6.3 Response of Renourished Beaches to Climate Change and a Rising Sea Level

Predictions of global warming and a world-wide sea level rise should now be taken into account in long-term planning for coastal management, including beach renourishment (Bird 1993). In the past century global average sea level has risen 10–20 cm, as measured by tide gauges located around the world (Church et al. 2001), as has the rate of rise (Mitchum et al. 2010). The latest (fifth) Assessment Report of the Intergovernmental Panel on Climate Change (IPCC 2013) estimates that sea level will continue to rise to the end of the 21st century and beyond, with estimates varying depending on scenarios for global emissions, because of thermal expansion of the oceans, melting of glaciers and ice sheets and changes in terrestrial water storage (Walsh et al. 2004).

Projected sea level rise has significant spatial variation. In Britain rates of sea level rise are influenced by relative land level changes: in Scotland sea level rise is somewhat reduced by the isostatic rebound (rising) of land since the last glacial period, while in the south of England relative sea level rise is expected to be greater because of land subsidence.

The effects of climate change may also include regional changes to storm frequency and intensity (IPCC 2013), which will impact on beaches. Small changes in wave direction may modify longshore drift and beach orientation.

The response of beaches to a rising sea level depends on a variety of factors, including geology, oceanographic and nearshore processes (wave climate, storm frequency), sediment supply and the intensity of hinterland development. Studies of the effects of a rising sea level on beach-fringed coasts have shown that erosion will be initiated or intensified as submergence proceeds, except where there is a continuing natural or artificial supply of sediment to maintain beaches at progressively higher levels. As the majority of beaches are already eroding (Bird 1985), the implications of global climate change will only serve to exacerbate erosion. The rate of beach erosion is expected to be two orders of magnitude greater than the increase in sea level (Leatherman 2001).

Bruun (1962, 1983) suggested that a sea level rise would cause sand to be eroded from the top of the beach and deposited offshore. This concept is known as the Bruun Rule, and is a widely used method of estimating the response of beaches to a rise in sea level. It indicates that a beach profile will maintain its shape as sea level rises, assuming it is able to migrate into an undeveloped hinterland. The concept is highly simplified and ignores the effects of longshore drift.

The coastline can be maintained by building sea walls and other protective structures to prevent erosion and submergence, but as has been noted, these are likely to cause further beach erosion, and in due course beaches (including those that have been renourished) will disappear, leaving an artificial coastline. It will be possible to maintain beaches by continuing renourishment as sea level rises, the limiting factors being the availability and cost of suitable renourishment sediment and the extent of hinterland submergence, which may have to be offset by building sea walls or raising coastal lowlands by landfill.

In Sydney, Australia, AECOM (2010) estimated that for an assumed increase of 0.1 m in sea level over the next decade, 9 million m^3 of sediment would be required to reinstate and maintain beach amenity and provide some storm protection. Subsequent annual renourishment was estimated to require a further 3 million m^3. In South Australia the Coast Protection Board has considered whether the Adelaide beaches (Sect. 4.3.4, p. 67) could be maintained if sea level rises as forecast, and found that renourishment would continue to be feasible, especially if inland sources of sand were used, until the sea has risen 20–30 cm above its present level, but thereafter it may be necessary to construct major sea walls and accept that the beaches will disappear (Wynne 1984). Inevitably, there will be future cost increases of beach renourishment projects, with more frequent and more substantial filling. As Weggel (1986) remarked 'if projections of an increasing rate of sea level rise are correct, it will become increasingly difficult to economically justify future beach renourishment projects'.

Planning for projected sea-level rise increases should be based on credible science, engineering and economics to ensure careful consideration of cost-effective methods of sustaining the coast (Williams 2013). Coastal managers may have to decide whether beach renourishment is a sustainable method of long-term coastal management. Coastline retreat, by moving structures landward or elevating them on pilings, or abandoning land and structures, may become necessary (Yohe et al. 1996).

References

Adriaanse LA, Coosen J (1991) Beach and dune nourishment and environmental aspects. Coast Eng 16:129–146

AECOM (2010) Beach sand nourishment scoping study: maintaining Sydney's beach amenity against climate change sea level rise. Prepared for Sydney Coastal Councils Group Inc (online)

Baca BJ, Lankford TE (1988) Myrtle beach nourishment project: biological monitoring. Report to City of Myrtle Beach, South Carolina

Barrett MG (1989) What is coastal management? In: Barrett MG (ed) Coastal management. Proceedings of the conference organised by the institution of civil engineering in Bournemouth, May 1989. Thomas Telford, London, pp 1–9

Basinski T (1994) Protection of Hel Peninsula. In: Rotnicki K (ed) Changes of the polish coastal zone. Quaternary Research Institute, Adam Mickiewicz University, Poznan, pp 53–56

Benedet L, Finkl CW, Dobrochinski JPH (2013) Optimization of nearshore dredge pit design to reduce impacts on adjacent beaches. J Coastal Res 29:519–525

Bird ECF (1985) Coastline changes: a global review. Wiley, Chichester

Bird ECF (1993) Submerging coasts: the effects of a rising sea level on coastal environments. Wiley, Chichester

Black DW, Donnelley LP, Settle RF (1988) An economic analysis of beach renourishment for the State of Delaware. University of Delaware

Bodge KR (1991) Damage benefits and cost sharing for shore protection projects. Shore Beach 59:11–18

Bourman RP (1990) Artificial beach progradation by quarry waste disposal at Rapid Bay South Australia. J Coastal Res 6:69–76 (special issue)

Brown AC, McLachlan A (1990) Ecology of sandy shores. Elsevier, Amsterdam

Bruun P (1962) Sea-level rise as a cause of shore erosion. J Waterw Harbour Div 88:117–130

Bruun P (1983) Review of conditions for the use of the Bruun rule of erosion. Coast Eng 7:77–89

Campbell TJ, Benedet L (2006) Beach nourishment magnitudes and trends in the U.S. J Coastal Res 39:57–64 (special issue)

Church JA, Gregory JM, Huybrechts P, Kuhn M, Lambeck K, Nhuan MT, Qin D, Woodworth PL (2001) Changes in sea level. In: Houghton JT et al (eds) Climate change 2001: the scientific basis. Cambridge University Press, Cambridge, pp 639–694

CIRIA (2010) Beach management manual (report no. C685). CIRIA, London

Clarke J, Brookes S (2008) Practical aspects of executing renourishment schemes on mixed beaches. Science report SCH00608B0FD-E-P (online), Environment Agency

Cooper K, Boyd S, Aldridge J, Rees H (2007) Cumulative impacts of aggregate extraction on seabed macro-invertebrate communities in an area off the east coast of the United Kingdom. J Sea Res 57:288–302

Dean RG (1987) Additional sediment inputs to the nearshore region. Shore Beach 55:76–81

Dean RG (2002) Beach nourishment: theory and practice. Advanced series on ocean engineering, vol 18. World Scientific, Singapore, p 399

Dette HH (1977) Effectiveness of beach deposit nourishment. In: Proceedings coastal sediments '77, pp 211–227

Dette HH (1990) Offshore sand dredge and delivery systems. Proc. 3rd Annual National Beach Preservation Technology Conference: 378–393

Finkl CW (2002) Long-term analysis of trends in shore protection based on papers appearing in the Journal of Coastal Research, 1984–2000. J Coastal Res 18:211–224

Goldberg J, Wilkinson C (2004) Global threats to coral reefs: coral bleaching, global climate change, disease, predator plagues, and invasive species. In: Wilkinson C (ed) Status of the coral reefs of the world. Australian Institute of Marine Science, Townsville, pp 67–92

Gopalakrishnan S, Smith MD, Slott JM, Murray AB (2011) The value of disappearing beaches: a hedonic pricing model with endogenous beach width. J Environ Econ Manage 61:297–310

Hamm L, Capobianco M, Dette HH, Lechuga A, Spanoff R, Stive MJF (2002) A summary of European experience with shore nourishment. Coast Eng 47:237–264
Herrera A, Gomez-Pina G, Fages L, De La Casa A, Muñoz-Perez JJ (2010) Environmental impact of beach nourishment: A case study of the Rio San Pedro Beach (SW Spain). Open Oceanogr J 4:32–41
Hoagland P, Jin D, Kite-Powell HL (2012) The costs of beach replenishment along the U.S. Atlantic Coast. J Coastal Res 28:199–204
Hobson RD (1977) Sediment handling and beach fill design. In: Proceedings coastal sediments '77, pp 167–180
IPCC (Intergovernmental Panel on Climate Change) (2013) Climate change 2013: the physical science basis. Contribution of working group I to the fifth assessment report of the Intergovernmental Panel on Climate Change, Geneva
Jordan LKB, Banks KW, Fisher LE, Walker BK, Gilliam DS (2010) Elevated sedimentation on coral reefs adjacent to a beach nourishment project. Mar Pollut Bull 60:261–271
La Porta B, Targusi M, Lattanzi L, La Valle P, Paganelli D, Nicoletti L (2009) Relict sand dredging for beach nourishment in the central Tyrrhenian Sea (Italy): effects on benthic assemblages. Mar Ecol 30(Suppl 1):97–104
Landry CE, Hindsley P (2011) Valuing beach quality with hedonic property models. Land Econ 87:92–108
Lankford TE, Baca BJ (1989) Comparative environmental impacts of various forms of beach nourishment. In: Proceedings coastal zone '89 conference, 2046–2059
Leatherman SP (2001) Social and economic costs of sea level rise. In: Douglas BC, Kearney MS, Leatherman SP (eds) Sea level rise, history and consequences. International geophysics series, vol 75. Academic Press, London, pp 181–224
Marsh GA, Turbeville DB (1981) The environmental impact of beach nourishment: two studies in south-eastern Florida. Shore Beach 47:40–44
Mitchum GT, Nerem RS, Merrifield MA, Gehrels WR (2010) Modern sea-level change estimates. In: Church JA, Woodworth PL, Aarup T, Wilson WS (eds) Sea-level rise and variability. Wiley, Chichester, pp 122–138
Nelson WG (1993) Beach restoration in the southeastern US: environmental effects and biological monitoring. Ocean Coast Manag 19:157–182
Neumann JE, Livesay ND (2001) Coastal structures, dynamic economic modeling. In: Mendelsohn R (ed) Global warming and the American economy, a regional assessment of climate change impacts. Edward Elgar, Cheltenham, pp 132–148
NOAA (National Oceanic and Atmospheric Administration) (2000) State, territory, and commonwealth beach nourishment programs: a national overview. Technical document 00-01, OCRM program policy series
Noriega R (2008) Environmental impacts of beach nourishment at Palm Beach, Gold Coast, Australia. Griffith Centre for Coastal Management, Research report 83. Southport, Queensland
Olsen EJ (1982) South Seas plantation beach improvement project. Shore Beach 50:6–10
Peterson C, Bishop MJ (2005) Assessing the environmental impacts of beach nourishment. Bioscience 55(10):887–896
Pezzuto PR, Resgalla JRC, Abreu JGN, Menezes JT (2006) Environmental impacts of nourishment of Balneario Camboriu Beach, SC, Brazil. J Coastal Res 39:863–868 (special issue)
Pilkey OH, Hume TM (2001) The shoreline erosion problem: lessons from the past. Water Atmos 9:22–23
Pitt A (2012) Beach nourishment for surfers and wider beaches. Paper presented at the 10th National Coast to Coast conference, Brisbane, September 2012 (online)
Pompe JJ, Rinehart JR (1994) Estimating the effect of wider beaches on coastal housing prices. Ocean Coast Manag 22:141–152
Pullen EJ, Naqui SM (1983) Biological impacts on beach replenishment and borrowing. Shore Beach 51(2):27–31

Rakocinski DF, Heard RW, Lecroy SE, Mclilland JA, Simons T (1996) Responses by macrobenthic assemblages to extensive beach restoration at Perdido key, Florida, USA. J Coastal Res 12(326):353

Reilly FJ, Bellis VJ (1983) The ecological impact of beach nourishment with dredged materials on the intertidal zone at Bogue Banks, North Carolina. U.S. Army Corps of Engineers, C.E.R.C, Fort Belvoir, Virginia. Miscellaneous paper 83–3

Roellig DA (1989) Shoreline response to beach nourishment. Proceedings coastal zone '89, pp 2104–2109

Rouch F, Bellessort B (1990) Man-made beaches more than 20 years on. In: Proceedings 22nd Coastal Engineering Conference, pp 2394–2401

Speybroeck J, Bonte D, Courtens W, Gheskiere T, Grootaert P, Maelfait J, Mathys M, Provoost S, Sabbe K, Stienen EWM, Lancker VV, Vincx M, Magda V, Degraer S (2006) Beach nourishment: an ecologically sound coastal defence alternative? A review. Aquat Conserv Mar Freshw Ecosyst 16:419–435

Stive M, Schipper M, Luijendijk A, de Vries JT, Aarninkhof S, Gelder-Maas C, de Vries ST, Henriquez M, Marx S, Ranasinghe R (2013) A new alternative to saving our beaches from sea-level rise: the sand engine. J Coastal Res 29:1001–1008

Stronge W (1990) Economic and fiscal impact of beach restoration: the case of Captivia. In: Proceedings of the 35th annual meeting, Florida Shore and Beach Preservation Association, Tallahassee, pp 74–86

Trembanis AC, Pilkey OH, Valverde HR (1999) Comparison of beach nourishment along the US, Atlantic, Great Lakes, Gulf of Mexico, and New England shorelines. Coast Manag 27:329–340

Van Dolah RF, Wendt PH, Martore RM, Levisen MV, Roumillat WA (1992) Final report for a physical and biological monitoring study of the Hilton Head beach nourishment project. Marine Resource Research Institute, Hilton Head Island

Van Dolah RF, Martore RM, Lynch AE, Levisen MV, Wendt PH, Whitaker DJ, Anderson WD (1994) Environmental evaluation of the Folly Beach nourishment project. US Army Corps of Engineers, Charleston

Walsh KJE, Betts H, Church J, Pittock AB, McInnes KL, Jackett DR, McDougall TJ (2004) Using sea level rise projections for urban planning in Australia. J Coastal Res 20:586–598

Watson JE (1973) The impact of the ethane pipeline on the marine ecosystems of Port Phillip Bay, Victoria. Victorian Nat 90:60–65

Weggel JR (1986) Economics of beach nourishment under scenario of rising sea level. J Waterw Port Coast Ocean Eng 112:418–427

Wellard K, Rimington N (2013) Beach-nourishment in Wales: practical issues, multiple benefits and ecological effects. In: Proceedings of the 2013 Institute of Civil Engineers Coasts, Marine Structures and Breakwaters Conference, Edinburgh, UK

Wilber DG, Clarke DG, Burlas MH (2006) Suspended sediment concentrations associated with a beach nourishment project on the northern coast of New Jersey. J Coastal Res 22:1035–1042

Williams SJ (2013) Sea-level rise implications for coastal regions. J Coastal Res 63:184–196 (special issue)

Wong PP (1985) Artificial coastlines: the example of Singapore. Z Geomorphol Supplementband 57:175–192

Wynne AA (1984) Adelaide coast protection strategy review. Coast Protection Board, South Australia

Yohe G, Neuman J, Marshall P, Ameden H (1996) The economic cost of greenhouse-induced sea-level rise for developed property in the United States. Clim Change 32:387–410

Chapter 7
Review of International Practices

Abstract Beach renourishment methods, magnitudes, design and evaluation procedures differ greatly from one country to the next, and there are variations in the level of research and available documentation. This Chapter provides a brief review of international practices and summarises trends in various countries.

Since the early beach renourishment projects in the 1920s on the coast of the United States, the technique has grown in popularity, becoming widely used for improving recreational beaches and protecting against coastal erosion and flooding. Beach renourishment has been widely applied in Western Europe, Australia and the United States, and in the last two decades its use has increased throughout the world. This review summarises some trends in various countries.

7.1 United States

The United States has had more beach renourishment projects than any other country. Beach renourishment remains the most widely utilised method for coastal protection. This is partly due to the fact that erosion has been prevalent, estimated to be occurring on 90 % of US beaches (Leatherman 1988). Reviews of national and regional practices include Leonard et al. (1990), Trembanis et al. (1999), Finkl et al. (2006), Campbell and Benedet (2006).

Although beach renourishment may have been used at San Pedro in southern California as early as 1919 (Herron 1980), the first well-documented project was in 1922 near the site of the amusement park at Coney Island, New York (Sect. 4.2.3, p. 46). Another early example is Waikiki Beach, Hawaii, which was renourished in 1939 as a recreational beach (Sect 4.2.8, p. 51). Beach renourishment soon became fashionable, and by 1991 over 640 km of beaches on the coastline of the United States had been renourished at a total cost of about $US8 billion (Davison et al. 1992).

E. Bird and N. Lewis, *Beach Renourishment*, SpringerBriefs in Earth Sciences,
DOI 10.1007/978-3-319-09728-2_7

It has been estimated that the amount of sediment placed on United States beaches by all forms of nourishment is of the order of half a billion m^3 (Trembanis et al. 1999; Campbell and Benedet 2006). By 2006 there were more than 200 renourished areas, and since the Coney Island project in the 1920s 9 million m^3 of sediment per year has been deposited on United States beaches (Campbell and Benedet 2006).

Beach renourishment has been most extensive on the barrier islands of the east coast of the United States (Trembanis et al. 1999), where by 1999 there had been 573 episodes of beach renourishment on 154 barrier island beaches (Valverde et al. 1999). In 1995 Pilkey had suggested that all the major coastal communities of the Atlantic seaboard of the United States coast had nourished beaches, or soon would have (Pilkey 1995).

One successful beach renourishment project was that on Miami Beach (Sect. 4.3.8, p. 73, 74), where the renourished beach provided a recreational amenity that revitalised Miami's international tourist industry (Finkl 1981). The project remains the most durable of any beach renourishment project in the United States (Finkl and Walker 2004).

States with severe beach erosion tend to have more comprehensive approaches to beach nourishment projects with established policies and funding mechanisms, although the majority of coastal states have at least some form of beach nourishment policy (NOAA 2000). An example is the Massachusetts Department of Environmental Protection's 'Beach Nourishment: Guide to Best Management Practices for Projects in Massachusetts' (MassDEP 2007).

The majority of renourishment works have been federally funded for the purposes of erosion control and flood mitigation, but some also undertaken for recreation purposes and for the beneficial use of dredged sediment from navigation channels. The federally funded processes have a standardised design that follows the guidelines in the Coastal Engineering Manual (USACE 2002), while non-federal funded projects are constructed by local governments and private owners, often with differing designs (Campbell and Benedet 2006).

Initial construction volumes of beach nourishment projects in the US range from 150 to 600 m^3 per metre (Campbell and Benedet 2006). Many are periodic, either to offset continuing erosion following renourishment, or as part of a long-term strategy. Delray Beach on the south east coast of Florida has been periodically renourished with five episodes of sand placement in 20 years (Hartog et al. 2008).

7.2 Australia and New Zealand

Beach erosion has become widespread in Australia, and in many places state and local government agencies have introduced beach renourishment projects. These have been principally for the protection of public infrastructure, private property and beach amenity, and are generally undertaken as part of coastal management plans. Eroding beaches have been renourished, particularly in coastal urban centres, notably Adelaide (Sect. 4.3.4, p. 67), the Gold Coast (Sect. 1, p. 3, Sect. 4.2.6, p. 48, Sect. 4.3.3, p. 66, Sect. 4.3.5, p. 70, Sect. 4.5, p. 86, Sect. 6.1.3, p. 111) and around Port Phillip Bay in Victoria (Sect. 2.7, p. 18, 19, Sect. 2.8, p. 20, Sect. 3, p. 29, Sect. 4.1, p. 42, 44, Sect. 4.2.1, p. 45, Sect. 4.2.7, p. 49, 50, Sect. 4.3.1, p. 56, Sect. 4.3.2, p. 64, Sect. 4.3.6, p. 71, Sect. 4.3.10, p. 75, 76, Sect. 4.4.4, p. 83, Sect. 4.5, p. 85, Sect. 5, p. 102–105, Sect. 6.1.1, p. 110, Sect. 6.2, p. 115, 116).

A review of beach renourishment practices in Australia was provided by Cooke et al. (2012), who identified 130 beaches that were renourished between 2001 and 2010. When compared with beach renourishment projects elsewhere, most Australian projects are small in scale (typically less than 50,000 m^3) but frequent (typically at intervals of less than a year). Sand is usually placed on the foreshore (i.e. between the high and low tide lines), possibly because of the importance of many Australian beaches for recreation. Mention has been made of Ettalong Beach in Broken Bay, 50 km north of Sydney (Sect. 4.3.4, p. 68), where small periodic renourishments (less than 50,000 m^3) are undertaken, the latest being in summer 2013 where sediment was mounded on the sub-aerial beach (Fig. 6.1). Large renourishment projects in Australia have included Lady Robinsons Beach close to Sydney Airport in Botany Bay, where in 1997 the beach volume was increased by placing 150,000 m^3 of sand and eight groynes, followed by 310,000 m^3 and five groynes in 2004–05 (AECOM 2010).

Beach renourishment has been widely applied in Port Phillip Bay in south-eastern Australia, a marine embayment with a coastline 256 km long. Between 1975 and 1990 some 20 of its beaches, averaging 1 km in length, had been renourished (Bird 1990). By 2010 this number had risen to 30 (Bird 2011), as shown in Fig. 7.1.

The results of beach renourishment in Port Phillip Bay have been favourable, this being a generally good environment for such projects, and most of the beaches have persisted for several years (Bird 1991). A review of beach renourishment projects in 2001 found that 'the majority of the beaches that have been renourished are in nearly as good a condition as when they were first constructed. Only a few need topping up with sand and only four needed to be rebuilt' (Vantree 2001). Nevertheless, periodic renourishment has been necessary, for example, on the north coast of Port Phillip Bay, where a sandy beach extends east from Station Pier, Port Melbourne to St Kilda, in front of a sea wall along Beach Street and Beaconsfield Parade. Mackenzie (1939) reported that sea walls and groynes had been constructed from 1898 onwards to halt coastline recession. The beach had been depleted during storms (large quantities of sand were swept from the beach across Beaconsfield Parade during the November 1934 storm surge), and parts have been artificially renourished, including a 900 m sector at Middle Park in 1976 and 2001. Subsequent storms, notably in February 2005, have depleted the beach and necessitated supplementary renourishment (Bird 2011).

Other notable locations of renourishment projects in Australia have included numerous along the east coast, such as those on the beaches of Townsville, along the Gold Coast, and at Port Stephens, Bate Bay, Collaroy-Narrabeen (Sect. 4.2.4, p. 47), Coffs Harbour, Towra, Silver Beach and Noosa (Sect. 4.3.4, p. 68, Sect. 4.3.8, p. 74).

Australia has a number of permanent by-passing projects, such as those at Noosa (Nankervis 2005) and Tweed Heads (Acworth and Lawson 2011; Boswood et al. 2001, 2005). These projects use permanent structures to transfer sediment by hydraulic pumping from one part of the beach to another (Noosa: Sect. 4.3.4, p. 67, Sect. 4.3.8, p. 74) or around (by-passing) structures at tidal inlets (Tweed River: Sect. 4.3.3, p. 66).

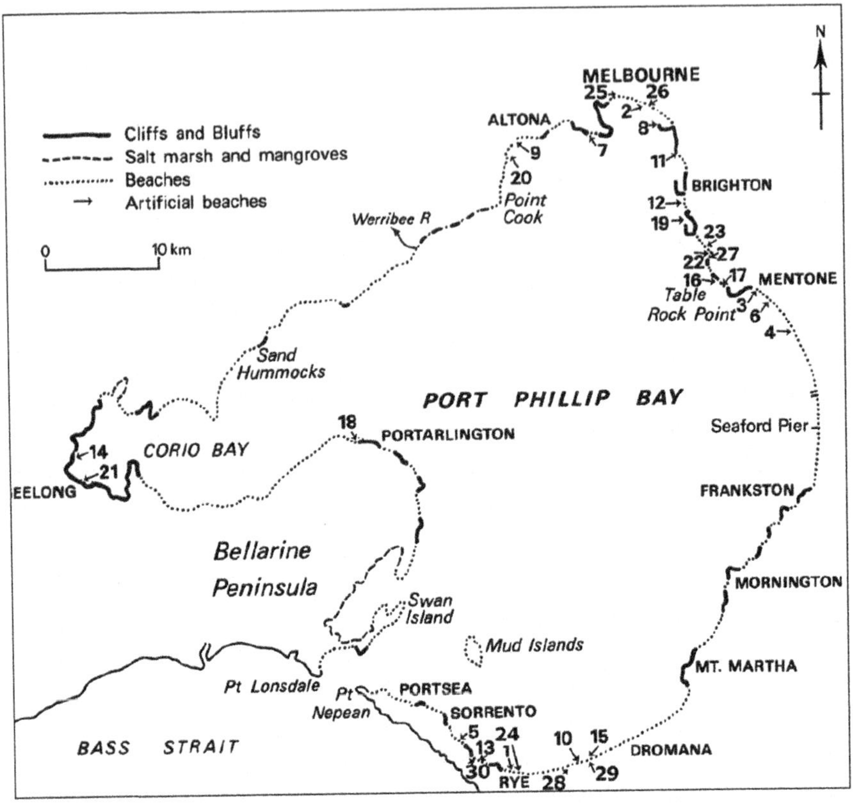

Fig. 7.1 Beach renourishment projects on the coast of Port Phillip Bay. *1* Rye (1975), *2* Middle Park (1976), *3* Mentone (1976), *4* Aspendale (1979), *5* Sorrento (1980), *6* Parkdale (1961), *7* Williamtown (1982), *8* St. Kilda (1982), *9* Altona (1982), *10* West Rosebud (1982), *11* Elwood (1983), *12* Brighton Park Street (1984), *13* Blairgownie (1984), *14* Geelong (1984), *15* Rosebud (1985), *16* Sandringham Quiet Corner (1986), *17* Watkins Bay (1986), *18* Portarlington (1986), *19* Brighton New Street (1987), *20* Altona South (1990), *21* Geelong, Eastern Beach (1990), *22* Sandringham, Edward Street (1993), *23* Hampton (1997), *24* Rye (1999), *25* Sandridge (1999), *26* Middle Park (2001), *27* Sandringham Royal Parade (2009), *28* Rye (2010), *29* Rosebud (2010), *30* Blairgowrie (2010)

A key limitation at a number of locations in Australia is the availability of suitable sediment for beach renourishment. As a result a number of studies have been carried out to establish the need for availability of offshore sources around Sydney (AECOM 2010) in relation to the estimated future sediment requirement for beach renourishment to protect Sydney beaches from continued erosion and future sea level rise. A significant limitation is that offshore resources of sand in New South Wales are currently protected by legislation, restricting the extraction of sand resources from the seabed.

In New Zealand a number of examples of beach renourishment projects exist, including at Tauranga Harbour (Sect. 4.2.3, p. 46, Sect. 4.4.2, p. 81) (De Lange and Healy 1990), Mission Bay and Kohimarama (Sect. 4.4.4, p. 83) (Papps and

Priestley 2005), and at the Bay of Plenty (Spiers and Healy 2005). These have been undertaken in an ad hoc manner to address issues such as reduced amenity and the undermining of seawalls. Sources of renourishment material are often offshore or dredged spoil from harbours or inlets.

7.3 Mainland Europe

Beach renourishment is practiced widely throughout mainland Europe, but is particularly common in the Netherlands, Spain, Belgium, Denmark, Germany, Italy and France. Reviews of European beach renourishment have been undertaken by Hamm et al. (2002) and Hanson et al. (2002). Early European renourishment projects were in the 1950s in Portugal, United Kingdom (discussed later) and Germany, followed by France in the early 1960s, Belgium and Italy in the late 1960s, the Netherlands and Denmark in the early 1970s and Spain in the 1980s.

Notable beach renourishment projects have included those on the islands of Sylt (Sect. 4.3.2, p. 62) and Nordeney (Sect. 4.4.2, p. 80) in Germany (Kunz 1990; Kelletat 1992; Hamm et al. 2002), near Ostend (Sect. 4.3.1, p. 56) in Belgium (Kerckaert et al. 1986), at Brest (Hallégouet and Guilcher 1990), at Saintes-Maries-de-la-Mer and Prado, near Marseille (Sect. 6.1.1, p. 109) in France (Rouch and Bellessort 1990), at Lonstrup in Denmark (Laustrup and Madsen 1994), at Lido di Jesolo near Venice (Sect. 4.2.4, p. 47) (Zunica 1990), Monte Circeo (Sect. 4.3.1, p. 59) and Marina di Cecina (Sect. 4.3.10, p. 78) in Italy (Ciprani et al. 1992; Evangelista et al. 1992) and at Rio San Pedro in Spain (Herrera et al. 2010).

Beach renourishment projects in countries applying the technique on a smaller scale include at Praia da Rocha (Sect. 4.3.1, p. 60) in Portugal (Psuty and Moreira 1990; Psuty et al. 1992), in Greece at Theologos on Rhodes, Sweden (Hanson et al. 2002), on the Polish Baltic coast (Rotnicki 1994), at Rosslare Strand and Donabate Beach in Ireland (Hanson et al. 2002), at St. George's Beach (Sect. 4.2.8, p. 51, Sect. 4.4.3, p. 82, Sect. 4.7, p. 92) in Malta (Firman et al. 2011), at Pirita Beach in Tallinn Bay, Estonia (Soomere and Kask 2006), at Odessa in the Ukraine (Shuisky 1994) and on the coast of Georgia (Zenkovich and Schwartz 1987).

The magnitude of renourishment and typical rates and volumes differ markedly between countries. Annual beach renourishment volumes in Europe are approximately 28 million m^3 (Hamm et al. 2002). Spain and the Netherlands continue to undertake the greatest number of beach renourishment projects in Europe. In Spain, although beach renourishment has only been widely practiced since 1983, there have been over 400 renourishment sites, mainly along the Mediterranean coast, while in the Netherlands over 200 renourishment projects have been undertaken since the early 1970s (Hanson et al. 2002). Renourishment projects in Spain, Italy and France tend to consist of smaller volumes spread over a greater number of sites, whilst in other countries such as Denmark and the Netherlands for example volume per site is much greater, typically over a million m^3.

Strategies and methods used in beach renourishment in Europe also differ. A comparison of national practices and policies by Hanson et al. (2002)

revealed differences linked to the main coastal problems of each country. In the Netherlands, Denmark, the United Kingdom and Germany beach renourishment projects typically form part of long-term intervention against coastal erosion and protection against flooding. Most notably, in the Netherlands the impetus has been to reduce the risk of flooding as a significant proportion of the land is below mean sea level. By contrast, in Spain, Italy, Portugal and France renourishment projects are driven strongly by recreational considerations and the need for dry beach width, which partly reflects the economic importance of recreation. In Spain renourishment has been used principally to protect property and sustain recreational beach space.

In the Netherlands, Germany, United Kingdom and Denmark the need for coastal protection against flooding has given rise to long-term beach renourishment and monitoring strategies (Hamm et al. 2002). The Netherlands in particular have long-term monitoring programs, with records of the position of dune foot and mean high and low water levels going back until 1840 (Bird 1996). After successful beach renourishment projects in the 1950s, Dutch government policy has been to maintain the coastline at its 1990 position, primarily using sand renourishment projects to prevent coastal recession. The Netherlands has the most strategic long-term approach to beach renourishment, with rigid design and legislation, most commonly using the 'Dutch Method' (Sect. 4.4.1, p. 79) but also experimenting with localised large-scale beach renourishment (Sect. 4.3.2, p. 62) (Royal HaskoningDHV 2009; Stive et al. 2013). The New Delta Committee, formed in 2007 to provide advice on the country's preparedness for mitigating flood risk attributable to accelerated sea level rise in the 21st century, the annual sand renourishment volume for the Dutch coast should increase from around 12 to 80 million m^3/year (Kabat et al. 2009).

Spain has used predominantly subaerial placement to increase dry beach width, often without sea walls or groynes, while Italy, Portugal and France commonly use such supporting structures. In Italy during the past 20 years about a hundred beaches have been artificially renourished (Valloni and Barsanti 2007). The many small-volume projects (40,000–100,000 m^3) are accompanied by several larger scale beach renourishments, as at Pellestrina (4 million m^3) and Cavallino (2 million m^3) near Venice, in the south of Italy at Paola (1 million m^3) and on the west coast at Ostia near Rome (1 million m^3). Borrow material is commonly sand and gravel and is almost exclusively from the sea floor (Barsanti et al. 2011). In Portugal the first beach renourishment was in 1950 at Estoril, near Lisbon, with the deposition of 15,000 m^3 of sand (Hanson et al. 2002). Despite this early use of the technique, coastal protection in Portugal has often been based on hard engineering structures (sea walls and groynes), generally built in response to emergency situations. Beach renourishment has only been used on relatively sheltered beaches, particularly on the Algarve coast, and generally with groynes to prevent downdrift losses. An example is Praia da Rocha (Fig. 4.18) (Psuty and Moreira 1990; Psuty et al. 1992). At Vale Do Lobo on the Algarve coast, erosion resulting from updrift construction of groyne fields and marina structures at Quarteira and Villa Moura was offset by beach renourishment in 1998–1999, with deposition of 600,000 m^3

of sand, pumped from offshore (Teixeira et al. 1998). Further renourishment was performed in 2006 (Teixeira 2009; Proenca et al. 2011). Other examples of renourishment projects in Portugal include those on the Algarve coast at Alvor, Cabanas Island and Cacela Peninsula (Dias et al. 2003); and in low-energy pocket and estuarine beaches on the west coast such as on the Tróia Peninsula (Silveira et al. 2011).

7.4 United Kingdom

Beach renourishment has been used in the United Kingdom since the 1950s as one of a number of methods for managing coastal defence (flooding and erosion). Coastal defence is primarily organised through management plans. Funding for most coastal defence projects, including beach renourishment, comes from the central government, quasi-governmental agencies and local councils. In some cases, and increasingly, financial contributions are sought by other stakeholders.

Beach renourishment is only one of a number of available responses to coastal problems in the United Kingdom, and there is usually an appraisal of possible solutions against a range of technical, economic, environmental and legislative criteria. Funding is often provided where the benefits of renourishment significantly outweigh the estimated costs. Other benefits such as improved opportunities for recreation and aesthetic values may influence the selection of an appropriate coastal defence scheme (Hanson et al. 2002). An example of a major beach renourishment scheme is that on the Lincolnshire coast between Skegness and Mablethorpe (Sect. 4.2.7, p. 49). In addition to the existing beach renourishment methods applied in the United Kingdom, the feasibility of large scale one-off renourishments, such as the 'sand engine' approach (Sect. 4.3.2, p. 62) is also been explored.

The success of beach renourishment projects in the United Kingdom has led to an increasing need for sources of sand and gravel, which at present largely come from the continental shelf. As of 2002 annual renourishment volumes in the United Kingdom were around 4 million m^3. In 1996 it was estimated that the need for beach renourishment sediment between 1995 and 2015 would be of the order of 209 million m^3 (CIRIA 1996), however this figure has proved to be an overestimate. More recent work on behalf of the Crown Estate has produced an estimate of the order of 40–80 million m^3 to the year 2033 (Royal HaskoningDHV 2013).

7.5 Africa

An early example of beach renourishment in Africa was at the port of Durban, South Africa in 1953. Durban's history of beach protection including beach renourishment has largely revolved around efforts to effectively operate a port. Beach renourishment projects here have typically been in response to storm events. For example, Vetch's and Addington Beach was renourished in 1982, then in 2009 and 2010, the latter two in response to a significant storm event in 2007. Each phase in 2009 and 2010 contributed approximately 250,000 m^3 of sediment to the beaches. Offshore sediment was dredged and pumped ashore by a 900 mm diameter pipe 1.4 km long. An offshore borrow site that had previously been used to reclaim berths within the harbour was used (Corbella and Stretch 2012).

In Egypt, sand brought from the desert near Cairo has been used to renourish beaches at Alexandria (Fanos et al. 1995), and beaches have been renourished on other African coasts as at Lagos in Nigeria (Ibe et al. 1991) and Keta Lagoon, Ghana (Nairn et al. 1998). In The Gambia erosion of Kololi Beach, a large tourist resort, led to renourishment with 1 million m^3 of sand (Royal HaskoningDHV 2000, 2007).

7.6 Asia

In many Asian nations activities such as destruction of mangroves for prawn culture have caused or worsened erosion, while artificial structures have diminished the extent of natural coastline. In south-east Asia several beaches have been renourished in Malaysia, Singapore, China and Japan. In Japan coastal erosion has been countered by building extensive sea walls and dumping tetrapods, because the government required that the coastline be stabilised by hard structures as a 'permanent' solution (Nakayama et al. 1982). The trend is now towards soft engineering methods, with the average annual beach renourishment rate in Japan now over 5 million m^3 (Hanson et al. 2002). Beaches have been created on the artificial coastline at the head of Tokyo Bay and renourished in seaside resorts as at Niigaata, and on the Toban coast facing the Seto Inland Sea (Kadomatsu et al. 1991).

In Tokyo Bay beaches were a major recreational area until the late 1950s, then land reclamation and the spread of port and industrial facilities overran them to produce a coastline dominated by concrete sea walls. Beach renourishment in front of these walls began in the 1970s, using sand dredged from the bay floor, and by 1990 nine artificial beaches with a total length of 13 km had been formed as part of a series of intensively used coastal recreation parks (Koike 1990) (Fig. 7.2).

China has an 18,000 km long coastline, and the traditional method of responding to coastal erosion has been to construct sea walls. In many locations the natural coastline has been replaced by developed and reclaimed land. Since the late 1970s there has been increased sand quarrying from beaches to provide sand for urban development. The construction of large dams for water supply, flood protection and energy generation has led to the depletion of beaches and widespread erosion along the coast of China (Kuang et al. 2011).

The first major beach nourishment project was completed in 1994 at Dalian City on Xinghai Bay. There have since been more than 16 renourishment projects, as at Beidaihe, northeast of the Dai River (Sect. 4.7, p. 92), and more are planned for coastal cities (Cai et al. 2010). In Hong Kong the beach in Repulse Bay was

Fig. 7.2 Intensive recreational use of a renourished beach on the shore of Tokyo Bay, Japan. © Geostudies

depleted by sand mining during the 2nd World War, and in 1990 it was renourished by pumping sediment from the sea floor (Leatherman 1996).

Beaches have been renourished on both the west and east coasts of Peninsular Malaysia, usually in response to erosion problems and intended to restore the beach and its recreational value. Commonly undertaken by the Malaysian Government beach renourishment projects are targeted at historically popular beaches, including Taman Robina in Semerang Perai Utara, Pulau Pinang; Batu 4 in Port Dickson in Negeri Sembilan; Pantai Sabak in Kelantan; and Kuala Terengganu to Kuala Ibai in Terengganu. Methods and rates vary with each renourishment. For example, at Teluk Cempedak in Kuantan on the east coast of Malaysia a 900 m length of beach fronting hotels was renourished with 176.000 m^3 sand pumped from the sea floor in 2004 (Razak et al. 2013). In 1996, beach renourishment in Kelantan used 1.2 million m^3 of sand along a 2.1 km stretch of beach.

In Singapore beach renourishment has often been linked to the reclamation of coastal land, with some beaches constructed in front of hard structures, such as sea walls.

More recently beach renourishment projects have been undertaken in other parts of Asia, including Sri Lanka and India. A one kilometre stretch of coastline in Marawila, Sri Lanka has been renourished with sand pumped from the sea floor and retained by offshore breakwaters. Renourishment was undertaken to combat coastal erosion but also to create sandy beaches for tourism and fishing industries. Beach renourishment in India has taken place to reduce the impacts of erosion caused by the seaport at Ennore north of Chennai (Sect. 4.2.3, p. 46) (Ramana Murthy et al. 2008).

7.7 Central and South America

In Brazil beach renourishment has primarily taken place in response to erosion and loss of beach area, exacerbated by over-development of the coastal zone. Examples include Copacabana Beach (Sect. 4.2.3, p. 46) (Vera Cruz 1972), Balneario Camboriu Beach (Sect. 4.2.5, p. 47) (Pezzuto et al. 2006) and Alegre Beach in Santa Catarina (Finkl and Walker 2004). In these areas, the socio-economic importance of the beaches, particularly in tourist areas, is a key driver for beach renourishment. At Gravata Beach in Santa Catarina, Brazil, emergency beach renourishment was undertaken to protect infrastructure (Finkl and Walker 2004). In Cuba the beach at the seaside resort of Varadero was damaged by storms, and sand from a coral cay was brought to restore it (Sect. 4.2.8, p. 51).

7.8 Summary

Beach renourishment methods and evaluation procedures differ greatly from one country to the next, and there are variations in the level of research and available documentation. In some countries, such as the Netherlands and Germany, beach renourishment is used to prevent flooding and erosion and generally forms part of proactive long-term coastal management, while in others beach renourishment is used as a remedial measure, undertaken on an ad hoc basis. Common methods range from direct fill of the upper beach, as in Spain, to whole-profile renourishment, which is a typical method in the United States, to by-passing schemes, widely used in Australia and United States.

References

Acworth C, Lawson S (2011) The Tweed river entrance sand bypassing project ten years of managing operations in a highly variable coastal system. In: Proceedings of the 20th NSW coastal conference, Tweed Heads, pp 1–23

AECOM (2010) Beach sand nourishment scoping study: maintaining Sydney's beach amenity against climate change sea level rise. Prepared for Sydney Coastal Councils Group Inc [online]

Barsanti M, Calda N, Valloni R (2011) The Italian coasts: a natural laboratory for the quality evaluation of beach replenishments. J Coast Res 61:1–7

Bird ECF (1990) Artificial beach nourishment on the shores of Port Philip Bay, Australia. J Coast Res 6:55–68

Bird ECF (1991) Changes on artificial beaches in Port Phillip Bay, Australia. Shore Beach 59:19–27

Bird ECF (1996) Beach Management. Wiley, Chichester

Bird ECF (2011) Changes on the coastline of Port Phillip Bay. Victorian Government Department of Sustainability and Environment

Boswood P, Victory S, Lawson S (2001) Placement Strategy and monitoring of the Tweed river entrance sand bypassing project nourishment works. In: Coasts and Ports Conference 2001: Proceedings of the 15th Australasian coastal and ocean engineering conference, the 8th Australasian Port and Harbour Conference, Institute of Engineers, Gold Coast, Australia, pp 253–258

Boswood P, Voisey CJ, Victory SJ, Robinson DA, Dyson A R, Lawson SR (2005) Beach response to Tweed river entrance sand bypassing operations. In: Proceedings of the 2005 coasts and ports conference, Australia, pp 199–124

Cai F, Dean RG, Liu J (2010) Beach nourishment in China: status and prospects. In: Proceedings of 32nd conference on coastal engineering, Shanghai, China, paper no 32

Campbell TJ, Benedet L (2006) Beach nourishment magnitudes and trends in the U.S. J Coast Res Special Issue 39:57–64

Ciprani LE, Dreoni AM, Pranzini E (1992) Nearshore morphological and sedimentological evolution induced by beach restoration: a case study. Bollettino di Oceanologica Teorica ed Applicata 10:279–295

CIRIA (1996) Beach recharge materials—demands and resources. In: Humphreys B, Coates T, Watkiss M, Harrison D (eds) Report. Construction Industry Research and Information Association, London, p 154

Cooke BC, Jones AR, Goodwin ID, Bishop MJ (2012) Nourishment practices on Australian sandy beaches: a review. J Environ Manage 113:319–327

Corbella S, Stretch DD (2012) Coastal defences on the KwaZulu-Natal coast of South Africa: a review with particular reference to geotextiles. J S Afr Inst Civil Eng [online] 54:55–64

Davison AT, Nicholls RJ, Leatherman SP (1992) Beach nourishment as a coastal management tool: an annotated bibliography on developments associated with the artificial nourishment of beaches. J Coast Res 8:984–1022

De Lange WP, Healy TR (1990) Renourishment of a flood-tidal delta adjacent beach, Tauranga Harbour, New Zealand. J Coast Res 6:627–640

Dias JMA, Ferreira Ó, Matias A, Vila A, Sá-Pires C (2003) Evaluation of soft protection techniques in Barrier islands by monitoring programs: case studies from Ria Formosa (Algarve, Portugal). J Coast Res 35:117–131

Evangelista S, La Monica GB, Landini B (1992) Artificial beach nourishment using fine crushed limestone gravel at Terracina (Latium, Italy). Bollettino di Oceanologica Teorica ed Applicata 10:273–278

Fanos AM, Khafagy AA, Dean RG (1995) Protective works on the Nile delta coast. J Coast Res 11:516–528

Finkl CW (1981) Beach nourishment, a practical method of erosion control. Geo-Mar Lett 1:155–161

Finkl CW, Walker HJ (2004) Beach nourishment. In: Schwartz M (ed) The encyclopedia of coastal science. Kluwer, Dordrecht, pp 147–161

Finkl CW, Benedet L, Campbell T (2006) Beach nourishment experience in the United States: status and trends in the 20th century. Shore Beach 74:8–16

Firman K, Kemp J, Finch D, Mallia A, Sciortino J (2011) Designing a sustainable beach replenishment scheme for a site in Malta. J Coastal Res 61:36–43

Hallégouet B, Guilcher A (1990) Moulin Blanc artificial beach, Brest, Western Brittany, France. J Coast Res 6:17–20

Hamm L, Capobianco M, Dette HH, Lechuga A, Spanoff R, Stive MJF (2002) A summary of European experience with shore nourishment. Coast Eng 47:237–264

Hanson H, Brampton A, Capobianco M, Dette HH, Hamm L, Laustrup C, Lechuga A, Spanhoff R (2002) Beach nourishment projects, practices, and objectives—a European overview. Coast Eng 47:81–111

Hartog WM, Benedet L, Walstra DJR, Van Koningsveld M, Stive MJF, Finkl CW (2008) Mechanisms that influence the performance of beach nourishment; a case study in Delray Beach, Florida, USA. J Coast Res 24:1304–1319

Herrera A, Gomez-Pina G, Fages L, De La Casa A, Muñoz-Perez JJ (2010) Environmental Impact of Beach Nourishment: A Case Study of the Rio San Pedro Beach (SW Spain). Open Oceanogr J 4:32–41
Herron WJ (1980) Artificial beaches in southern California. Shore Beach 48:3–12
Ibe AC, Awosika LF, Ibe CE, Inegbedion LE (1991) Monitoring of the 1985–86 beach nourishment project at Bar Beach, Victoria Island, Lagos, Nigeria. In: Proceedings of Coastal Zone '91, pp 534–552
Kabat P, Fresco LO, Stive MJF, Veerman CP, van Alphen JSLJ, Parmet BWAM, Hazeleger W, Katsman CA (2009) Dutch coasts in transition. Nat Geosci 2:450–452
Kadomatsu T, Uda T, Fujiwara K (1991) Beach nourishment and field observations of beach changes on the Toban coast facing Seto Inland Sea. Mar Pollut Bull 23:155–159
Kelletat D (1992) Coastal erosion and protection measures at the German North Sea coast. J Coastal Res 8:699–711
Kerckaert P, Paul PL, Roovers AN, De Cand TP (1986) Artificial beach nourishment on Belgian east coast. J Waterways Port Coast Ocean Eng 112:560–571
Koike K (1990) Artificial beach construction on the shores of Tokyo Bay, Japan. J Coast Res 6:45–54
Kuang CP, Pan Y, Zhang Y, Liu SG, Yang YX, Zhang JB, Dong P (2011) Performance evaluation of a beach nourishment project at West Beach in Beidaihe, China. J Coast Res 27:769–783
Kunz H (1990) Artificial beach nourishment on Nordeney, a case study. In: Proceedings of 22nd coastal engineering conference, pp 3254–3267
Laustrup C, Madsen HT (1994) Design of breakwaters and beach nourishment. In: Proceedings of the 24th international coastal engineering conference, ASCE, pp 1359–1372
Leatherman SP (1988) Beach response to strategies to accelerated sea level rise. In: Proceedings of the 2nd North American conference on preparing for climate change. The Climate Institute, Washington, DC, pp 353–358
Leatherman SP (1996) Shoreline stabilisation approaches in response to sea level rise: U.S. experience and implications for Pacific Island and Asian nations. Water Air Soil Pollut 92:149–157
Leonard LA, Clayton T, Pilkey OH (1990) An analysis of replenished beach design parameters on U.S. East Coast barrier islands. J Coast Res 6:15–36
Mackenzie AD (1939) Coastal erosion in Victoria. Trans Inst Eng Aust 20:229–236
Massachusetts Department of Environmental Protection (MassDep) (2007) Beach Nourishment: Guide to Best Management Practices for Projects in Massachusetts [online]. Report produced for Massachusetts Department of Environmental Protection, Boston
Nairn RB, MacIntosh KJ, Hayes MO, Nai G, Anthonio SL, Valley WS (1998) Coastal erosion at Keta Lagoon, Ghana—large-scale solution to a large-scale problem. In: Proceedings of the 26th international conference on coastal engineering, ICCE-98, Copenhagen, Denmark, pp 3192–3205
Nakayama JP, Kataoka S, Obara K (1982) Construction of artificial beaches in Japan. Civil Eng Jpn 21:100–113
Nankervis L (2005) Beach nourishment with the submarine sandshifter. In: Proceedings of the 2005 coasts and ports conference, Australia, pp 341–344
National Oceanic and Atmospheric Administration (NOAA) (2000) State, territory, and commonwealth beach nourishment programs: a national overview. Technical Document 00-01, OCRM Program Policy Series
Papps D, Priestley S (2005) Design and performance of replenishment projects on Auckland's eastern beaches. In: Proceedings of the 2005 Coasts and Ports Conference, Australia, pp 125–130
Pezzuto PR, Resgalla JRC, Abreu JGN, Menezes JT (2006) Environmental impacts of nourishment of Balneario Camboriu Beach, SC, Brazil. J Coast Res 39:863–868
Pilkey OH (1995) The fox guarding the hen house. J Coast Res 11:iii–v
Proenca B, Oliveira FSBF, Sancho F (2011) Coastal erosion management in Algarve (Portugal)—a beach nourishment case study. J Coast Res 61:328–334

Psuty NP, Moreira ME (1990) Nourishment of a cliffed coastline, Praia da Rocha, the Algarve, Portugal. J Coast Res 6:21–32
Psuty NP, Eugine M, Moreira SA (1992) Characteristics and longevity of beach nourishment at Praia da Rocha, Portugal. J Coast Res 8:660–676
Ramana Murthy MV, Mani JS, Subramanian BR (2008) Evolution and performance of beach fill at Ennore Seaport, southeast coast of India. J Coastal Res, 24:232–243
Razak MSA, Roelvink D, Reyns J (2013) Beach response due to sand nourishment on the east coast of Malaysia. In: Proceedings of the ICE—maritime engineering, Dec 2013, pp 151–174
Rotnicki K (ed) (1994) Changes of the Polish Coastal Zone. Quaternary Research Institute, Adam Mickiewicz University, Poznan
Rouch F, Bellessort B (1990) Man-made beaches more than 20 years on. In: Proceedings of 22nd coastal engineering conference, pp 2394–2401
Royal HaskoningDHV (2000) Coastal protection study, feasibility report coastal protection study at Kololi Beach, The Gambia
Royal HaskoningDHV (2007) Coastal zone management handbook for The Gambia
Royal HaskoningDHV (2009) Sand engine, Environmental Impact Assessment, Holland
Royal HaskoningDHV (2013) The need for beach nourishment in England and Wales. Report produced for the Crown Estate
Shuisky YD (1994) An experience of studying artificial ground terraces as a means of coastal protection. Ocean Coast Manage 22:127–139
Silveira TM, Kraus NC, Psuty NP, Andrade F (2011) Beach nourishment on Tróia Peninsula, Portugal. J Coast Res 59:173–180
Soomere T, Kask, J (2006) Modelling of sand transport at Pirita Beach and formulation of principles of beach nourishment. In: Baltic Sea Geology, 9th Marine Geological Conference, Latvia, pp 93–95
Spiers KC, Healy TR (2005) Continued beach renourishment from dredge spoil disposal, Bay of Plenty, New Zealand. In: Proceedings of the 2005 coasts and ports conference, Australia, pp 363–368
Stive M, Schipper M, Luijendijk A, Theil de Vries J, Aarninkhof S, Gelder-Maas C, Theil de Vries S, M Henriquez, Marx S, Ranasinghe R (2013) A new alternative to saving our beaches from sea-level rise: the sand engine. J Coast Res 29:1001–1008
Teixeira SB (2009) Artificial beach nourishment with aggregates in the Algarve. Curso Técnico de Dragagens, Aveiro (in Portuguese)
Teixeira SB, Furtado A, Gaspar M (1998) Avaliação do Impacte da Alimentação Artificial da Praia de Vale do Lobo nas Comunidades de Bivalves (Algarve-Portugal). Seminário sobre Dragagens, Dragados e Ambientes Costeiros, Associação Eurocoast-Portugal, Porto, pp 47–56
Trembanis AC, Pilkey OH, Valverde HR (1999) Comparison of Beach nourishment along the US, Atlantic, Great Lakes, Gulf of Mexico, and New England shorelines. Coast Manage 27:329–340
US Army Corps of Engineers (USACE) (2002) Coastal Engineering Manual. Washington DC
Valloni R, Barsanti M (2007) Artificial Beach Nourishment Projects in Italy. Rapport Commission International Mer Mediterranee 38:706
Valverde RH, Trembanis AC, Pilkey OH (1999) Summary of beach nourishment episodes on the U.S. East Coast barrier islands. J Coast Res 15:1100–1118
Vantree (2001) Beaches at Risk. Vantree Pty Ltd, Caulfield
Vera Cruz D (1972) Artificial nourishment at Copacabana Beach. In: Proceedings of 13th coastal engineering conference, pp 1451–1463
Zenkovich VP, Schwartz ML (1987) Protecting the Black Sea-Georgian SSR gravel coast. J Coast Res 3:201–209
Zunica M (1990) Beach behaviour and defences along the Lido di Jesolo, Gulf of Venice, Italy. J Coastal Res 6:709–719

Chapter 8
Conclusion

Beach renourishment has been an effective way of maintaining beaches as a recreational resource and scenic amenity, and of countering coastal erosion in several countries around the world, and is widely preferred to the use of solid structures such as sea walls and groynes for coastal stabilisation. It is acknowledged that beach renourishment is not a permanent solution to the problem of beach erosion and coastal stabilisation, but nor is the use of artificial structures. Coastal management is necessarily a long-term strategy, with techniques likely to improve with experience.

A variety of approaches to beach renourishment have been developed to assist design, and a variety of monitoring techniques have been widely employed to evaluate performance. However, despite this experience, it remains difficult to predict the morphological behaviour of renourished beaches because of the complex interactions which occur between coastal processes, beach and nearshore morphology and sediments (Blott and Pye 2004).

Beach erosion is likely to increase in future because of a rising sea level and a possible increase in the frequency and severity of storms (Hoagland et al. 2012). In consequence, there will be an increase in the use of beach renourishment: Campbell and Benedet (2006) have suggested that beach nourishment volumes in the United States could double or triple over the next 25 years. One problem will be the securing of sufficient sources of sediment for beach renourishment, which will increase demands for sea floor dredging and for quarrying of sand and gravel for this purpose.

References

Blott SJ, Pye K (2004) Morphological and sedimentological changes on an artificially nourished beach, Lincolnshire, UK. J Coastal Res 20:214–233

Campbell TJ, Benedet L (2006) Beach nourishment magnitudes and trends in the U.S. J Coastal Res Spec Issue 39:57–64

Hoagland P, Jin D, Kite-Powell HL (2012) The costs of beach replenishment along the U.S. Atlantic Coast. J Coastal Res 28:199–204

E. Bird and N. Lewis, *Beach Renourishment*, SpringerBriefs in Earth Sciences,
DOI 10.1007/978-3-319-09728-2_8